Smart Coatings on Fibers and Textiles

Smart Coatings on Fibers and Textiles

Special Issue Editor

Mazeyar Parvinzadeh Gashti

MDPI • Basel • Beijing • Wuhan • Barcelona • Belgrade • Manchester • Tokyo • Cluj • Tianjin

Special Issue Editor
Mazeyar Parvinzadeh Gashti
PRE Labs Inc.
Canada

Editorial Office
MDPI
St. Alban-Anlage 66
4052 Basel, Switzerland

This is a reprint of articles from the Special Issue published online in the open access journal *Fibers* (ISSN 2079-6439) (available at: https://www.mdpi.com/journal/fibers/special_issues/smart_coatings_Fibers_textiles).

For citation purposes, cite each article independently as indicated on the article page online and as indicated below:

LastName, A.A.; LastName, B.B.; LastName, C.C. Article Title. *Journal Name* **Year**, *Article Number*, Page Range.

ISBN 978-3-03928-462-7 (Pbk)
ISBN 978-3-03928-463-4 (PDF)

© 2020 by the authors. Articles in this book are Open Access and distributed under the Creative Commons Attribution (CC BY) license, which allows users to download, copy and build upon published articles, as long as the author and publisher are properly credited, which ensures maximum dissemination and a wider impact of our publications.

The book as a whole is distributed by MDPI under the terms and conditions of the Creative Commons license CC BY-NC-ND.

Contents

About the Special Issue Editor vii

Preface to "Smart Coatings on Fibers and Textiles" ix

Chen Xue and Lee D. Wilson
A Spectroscopic Study of Solid-Phase Chitosan/Cyclodextrin-Based Electrospun Fibers
Reprinted from: *Fibers* **2019**, *7*, 48, doi:10.3390/fib7050048 1

Dariush Firouzi, Chan Y. Ching, Syed N. Rizvi and P. Ravi Selvaganapathy
Development of Oxygen-Plasma-Surface-Treated UHMWPE Fabric Coated with a Mixture of SiC/Polyurethane for Protection against Puncture and Needle Threats
Reprinted from: *Fibers* **2019**, *7*, 46, doi:10.3390/fib7050046 15

Noor Sanbhal, Xiakeer Saitaer, Mazhar Peerzada, Ali Habboush, Fujun Wang and Lu Wang
One-Step Surface Functionalized Hydrophilic Polypropylene Meshes for Hernia Repair Using Bio-Inspired Polydopamine
Reprinted from: *Fibers* **2019**, *7*, 6, doi:10.3390/fib7010006 29

Nongyi Cheng, Kwang-Won Park and Trisha L. Andrew
Solvent-Free Reactive Vapor Deposition for Functional Fabrics: Separating Oil–Water Mixtures with Fabrics
Reprinted from: *Fibers* **2019**, *7*, 2, doi:10.3390/fib7010002 39

Daiki Komoto, Ryoka Ikeda, Tetsuya Furuike and Hiroshi Tamura
Preparation of Chitosan-Coated Poly(L-Lactic Acid) Fibers for Suture Threads
Reprinted from: *Fibers* **2018**, *6*, 84, doi:10.3390/fib6040084 49

Mei-Ling Huang and Chien-Chang Fu
Applying Image Processing to the Textile Grading of Fleece Based on Pilling Assessment
Reprinted from: *Fibers* **2018**, *6*, 73, doi:10.3390/fib6040073 61

Lorenzo Scherino, Martino Giaquinto, Alberto Micco, Anna Aliberti, Eugenia Bobeico, Vera La Ferrara, Menotti Ruvo, Armando Ricciardi and Andrea Cusano
A Time-Efficient Dip Coating Technique for the Deposition of Microgels onto the Optical Fiber Tip
Reprinted from: *Fibers* **2018**, *6*, 72, doi:10.3390/fib6040072 75

Morgan Baima and Trisha L. Andrew
Fluoropolymer-Wrapped Conductive Threads for Textile Touch Sensors Operating via the Triboelectric Effect
Reprinted from: *Fibers* **2018**, *6*, 41, doi:10.3390/fib6020041 87

Kony Chatterjee, Jordan Tabor and Tushar K. Ghosh
Electrically Conductive Coatings for Fiber-Based E-Textiles
Reprinted from: *Fibers* **2019**, *7*, 51, doi:10.3390/fib7060051 95

Dharshika Kongahage and Javad Foroughi
Actuator Materials: Review on Recent Advances and Future Outlook for Smart Textiles
Reprinted from: *Fibers* **2019**, *7*, 21, doi:10.3390/fib7030021 141

Jiali Yu and Chi-Wai Kan
Review on Fabrication of Structurally Colored Fibers by Electrospinning
Reprinted from: *Fibers* **2018**, 6, 70, doi:10.3390/fib6040070 . . . . **165**

About the Special Issue Editor

Mazeyar Parvinzadeh Gashti got his Ph.D. degree in textile chemistry and fiber science in 2010. He served three Postdoctoral Fellowships at the University of Bern, Laval University and McGill University. He has held several positions with commercial businesses as Chief Scientist with responsibilities for product development and enhancement on polymers, composites, additives, fibers and textile materials. He also serves as member of the editorial board of several international journals on polymers, textiles and materials. Gashti has published 81 papers in different distinguished journals. His research expertise and interests are synthesis of bio- and nanomaterials, application of biomaterials in controlled drug delivery of composite particles, synthesis of conductive composite materials and process of smart textiles.

Preface to "Smart Coatings on Fibers and Textiles"

Today, we know that nanotechnology has been considered extensively in fiber and textile engineering in order to perform new functionalities. Ultrafine nanoparticles can transfer their intrinsic properties to fibers and textiles by surface coatings. Although several research studies confirmed such functionalities, research is still in progress in laboratories around the world to establish further results. Smart coatings can also be performed on textile products through other methods, such as plasma and laser coatings, sol-gel techniques, magnetron sputter coating, layer-by-layer techniques and crosslinking using polymers. Several properties are demonstrated using these methods, such as antibacterial, superhydrophobic, fire retardant, self–cleaning, super hydrophilic, moth-proofing, electromagnetic shielding, and electrical conductivity. In this Special Issue, original research papers, as well as reviews, are welcome. The goal is to gather contributions on various aspects related to smart coatings, including preparation, analyses, industrial uses, as well as their potential toxicity to humans during their usage. I hope that this Special Issue will provide the scientific community with a thorough overview of the current research on smart fibers and textiles.

Mazeyar Parvinzadeh Gashti
Special Issue Editor

Article

A Spectroscopic Study of Solid-Phase Chitosan/Cyclodextrin-Based Electrospun Fibers

Chen Xue and Lee D. Wilson *

Department of Chemistry, University of Saskatchewan, 110 Science Place, Saskatoon, SK S7N 5C9, Canada; chx257@mail.usask.ca
* Correspondence: lee.wilson@usask.ca; Tel.: +1-306-966-2961; Fax: +1-306-966-4730

Received: 1 April 2019; Accepted: 16 May 2019; Published: 22 May 2019

Abstract: In this study, chitosan (chi)/hydroxypropyl-β-cyclodextrin (HPCD) 2:20 and 2:50 Chi:HPCD fibers were assembled via an electrospinning process that contained a mixture of chitosan and HPCD with trifluoroacetic acid (TFA) as a solvent. Complementary thermal analysis (thermal gravimetric analysis (TGA)/differential scanning calorimetry (DSC)) and spectroscopic methods (Raman/IR/NMR) were used to evaluate the structure and composition of the fiber assemblies. This study highlights the multifunctional role of TFA as a solvent, proton donor and electrostatically bound pendant group to chitosan, where the formation of a ternary complex occurs via supramolecular host–guest interactions. This work contributes further insight on the formation and stability of such ternary (chitosan + HPCD + solvent) electrospun fibers and their potential utility as "smart" fiber coatings for advanced applications.

Keywords: cyclodextrin; chitosan; electrospinning; fiber; assembly

1. Introduction

Chitosan is a copolymer containing β-(1-4)-linked N-acetyl-D-glucosamine and D-glucosamine monomer units derived from the deacetylation of chitin, where the degree of acetylation depends on the extent of hydrolysis [1–7]. The amino polysaccharide units of chitosan contribute to its biodegradability and biocompatibility, along with its unique physicochemical properties relevant to adsorption. According to the degree of ionization of glucosamine groups of chitosan, the antimicrobial and adsorption properties can be altered by variable pH levels [8–11]. Diverse applications of chitosan have been found in many fields that exploit its pH-dependent binding properties, as evidenced in wastewater treatment, food preservation and biomedical devices [8,9,12–18]. To attain additional performance in these applications, research has also focused on the design of different morphological forms of chitosan that include nanofibrous systems, owing to the high surface area of these biomaterials [2–4,14,16–20]. Among the various approaches in the design of fibrous and nanofibrous materials, electrospinning methods have gained increasing attention [16–18,21–24]. While chitosan is soluble in acidic aqueous solution, technical difficulties related to the electrospinning of uniform fibers arise due to repulsive interactions between the cationic sub-units of chitosan that attenuate chain entanglement effects [4,19–21,25]. The judicious choice of additive components offers a solution to offset such charge repulsion effects during electrospinning [4,7,26–30]. For example, Burns et al. reported the use of trifluoroacetic acid (TFA) and hydroxypropyl β-cyclodextrin (HPCD) as additives to assist in the electrospinning of chitosan nanofibers [31]. The potential of HPCD to form noncovalent host–guest complexes with various molecular species, in conjunction with the polyelectrolyte nature of chitosan, may further enhance the utility of such nanofibrous materials as advanced coating materials [31,32]. Notwithstanding the complexation properties of HPCD, the molecular level details that account for the uniform formation of chitosan fibers via this pathway are not well understood. While the authors allude to the possibility that inclusion complex formation occurs between chitosan and HPCD, further insight is required to establish the origins of the improvement of fiber formation in this ternary (HPCD + chitosan + TFA) system to advance the field of chitosan nanofiber materials.

Herein, the overall goal of this study relates to an investigation of the structural role of ternary components (HPCD + chitosan + TFA) in the solid state via complementary spectroscopic techniques for chitosan-based fibers. The results of this study are foreseen to provide insight on the fiber formation process in such ternary component systems that will aide in the development of improved electrospinning formulations for chitosan systems. To address this goal, preparations of Chi:HPCD fibers at variable ratios were carried out via electrospinning in nonaqueous media. The composition and structural characterization of the electrospun fibers in the solid state was carried out using thermal analysis and spectroscopic (FT-IR and Raman) methods. Raman spectral imaging was assisted by a rhodamine dye probe to gain further structural information on the chitosan fiber composite materials reported herein.

2. Materials and Methods

2.1. Materials

Research grade 6-O-hydroxypropyl β-cyclodextrin (HPCD) (degree of substitution (DS) = 4.6), was purchased from CYCLOLAB, Ltd. (Budapest, Hungary) and was used as received. Rhodamine 6G was purchased from Allied Chemical (Morristown, New Jersey, NJ, USA) and used as received. Low molecular weight (LMW) chitosan (Chi, 75–80% deacetylation and molecular weights in a range of 50–190 kDa measured by Brookfield viscosity 20 cps), deuterated dimethyl sulfoxide (DMSO-d_6, 99.9%), tetrahydrofuran (THF) and trifluoroacetic acid (TFA) was bought from Sigma-Aldrich Canada Ltd. (Oakville, ON, Canada).

2.2. Solution Preparation

All solutions used for electrospinning were formulated according to Burns et al. [31], where the designated mass of chitosan and HPCD (wt%) was added to neat TFA and allowed to mix overnight with stirring (100 rpm) at 23 °C. All solutions were kept at 4 °C and consumed within 72 h.

2.3. Electrospinning

Solutions prepared in Section 2.2 (2.5 mL) were placed into a 10 mL syringe with a metal needle (Inner Diameter = 0.508 mm) that was operated by a Cole-Parmer 78-0100c syringe pump (Cole-Parmer, Montreal, QC, Canada). The distance between the needle tip and collector plate was set to 11.5 cm. The flow rate was controlled at 0.1 mL h^{-1}, where a high voltage was produced by a high voltage power supply (Spellman CZE 1000R) (Spellman, Hauppauge, NY, USA) and was applied between the needle tip and collector plate during electrospinning. The voltage was slowly adjusted from 7 to 15 kV to achieve a stable Taylor cone. The resulting fiber product was accumulated onto a foil-covered, electrically grounded collector plate. The electrospinning apparatus and molecular structures of precursors and solvent is shown in Scheme 1.

2.4. ^{1}H NMR Spectroscopy in Solution

The HPCD content of the fiber samples was estimated using a quantitative NMR (qNMR) method that was adapted from a previous report [33]. A 1% (w/w) THF/DMSO-d_6 solution was prepared by adding a desired amount of THF to DMSO-d_6 to which ca. 5 mg of the *as-spun* fiber was dissolved with ~600 mg of solvent (THF/DMSO-d_6) in a 5 mm NMR tube. ^{1}H NMR spectra were obtained using a wide-bore (89 mm) 11.7T (500 MHz) Oxford superconducting magnet system (Bruker BioSpin Corp., Billerica, MA, USA) equipped with a 5 mm Pa Tx1 probe. THF served as an internal quantitative standard for estimation of HPCD content of a fiber sample.

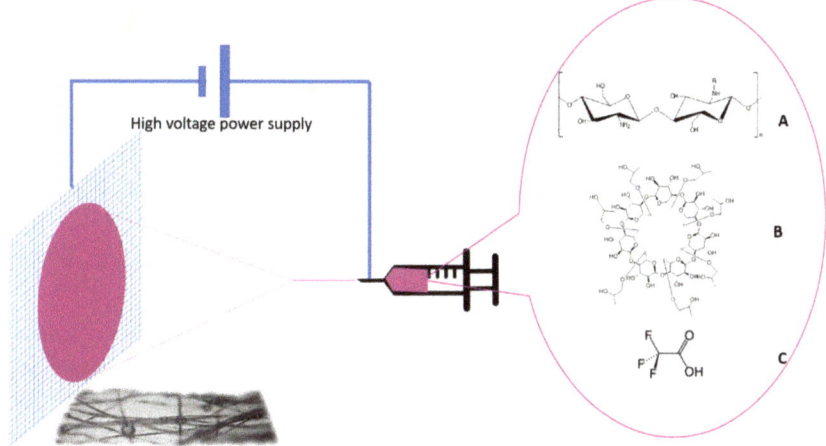

Scheme 1. A schematic illustration of the electrospinning setup and molecular structure of the precursors: (**A**) chitosan where R = –COCH$_3$ or H depending the degree of deacetylation, where n depends on the relative molecular weight of the biopolymer; (**B**) hydroxypropyl-β-cyclodextrin (HPCD); (**C**) trifluoroacetic acid (TFA) is the solvent system.

2.5. FT-IR Spectroscopy

Reflectance-based IR spectra were obtained with a Bio-RAD FTS-40 (Bio-RAD Laboratories, Inc., Hercules, CA, USA) instrument. IR samples were prepared by thoroughly mixing a sample (~5 mg) with FT-IR grade KBr (~50 mg) with a mortar and pestle. FT-IR spectra were measured at 23 °C with a resolution of 4 cm^{-1} over a spectral range of 400–4000 cm^{-1} (Kubelka-Munk intensity units) where the background spectrum of KBr was subtracted. Multiple scans (n = 16) were collected for better signal-to-noise ratio.

2.6. Raman Spectroscopy

A Renishaw InVia Reflex Raman microscope (785 nm solid state diode laser with a 1200 lines/mm grating system) (Renishaw plc, New Mills, UK) was used to collect One-dimensional (1–D) Raman spectra with a Pelletier cooled CCD (charge coupled device) detector (400 × 576 pixels). The instrument wavelength was calibrated at 520 cm^{-1} using an internal Si (110) sample.

2.7. Raman Imaging with Dye Probe

A procedure was adapted from a previous reported work for acquiring Raman spectra [34]. High resolution Raman 2-D (two-dimensional) spectral imaging (1.1 × 1.1 μm pixel size) was acquired using the instrument described above with a Leica 50× long-working-distance objective (numerical aperture = 0.50) using Streamline mode within the instrument software (Renishaw Wire V3.4) (Renishaw plc, New Mills, UK) with a 12 s exposure time under a static scan mode centered at 790 cm^{-1}. An effective spectral range of 596–985 cm^{-1} is achieved by the monochromator grating. 2-D images (92 × 70 spectra) were collected at specific Raman shifts in a uniform 101 × 77 μm grid to create Raman micro-images with spectral intensity (relative to baseline) for respective Raman shifts. Prior to creation of Raman images and to minimize any baseline sloping and offset effects that result from experimental artefacts, we performed the baseline linearization and normalization of the spectra for Raman imaging using the built in Renishaw Wire V3.4 method. The color intensity of the pixels on the image corresponded to the integration of the respective bands (610 and 850 cm^{-1}) to baseline as required. To highlight the chitosan spectral region, the results from dividing the band integrations at 610 cm^{-1} by the band at 850 cm^{-1} were used to construct the Raman images for chitosan. Prior to the Raman imaging, dried Chi:HPCD

2:50 sample was placed onto an Au-coated Si wafer that was then soaked with a Rhodamine 6G dye in a benzene solution (~0.1 μM), and finally dried under ambient ventilation for 24 h.

2.8. Differential Scanning Calorimetry (DSC)

A TA Q20 thermal analyzer (TA instruments, New Castle, DE, USA) was used to obtain DSC profiles of HPCD, physical mixtures of components, chitosan and 2%:50% fiber over the range between 40 °C and 190 °C. Solid samples were hermetically sealed in aluminum pans, where the sample weight varied from 1.55 mg to 2.60 mg before spectral acquisition. DSC profiles were recorded at a 10 °C/min scan rate under dry nitrogen gas with a flow rate at 50 mL/min.

2.9. Thermal Gravimetric Analysis (TGA)

A Q50 (TA instruments, New Castle, DE, USA) thermogravimetric analyzer was employed to obtain weight loss profiles. The experiments were executed under a nitrogen atmosphere with a heating rate (5 °C min^{-1}) up to a maximum temperature (500 °C). First derivative plots (weight %/°C vs. temperature (°C)) were generated to determine the thermal stability of materials.

2.10. Scanning Electron Microscopy (SEM)

Scanning electron microscopy (SEM) images with 508 dpi resolution were acquired with a JSM-6010LV microscope (JEOL, Ltd., Tokyo, Japan) at various magnifications (1000×, 5000× and 3000×). The samples were fixed onto a sample mounting stub with conductive carbon tape.

3. Results and Discussion

As indicated in the introduction section above, the role of additives (TFA, HPCD) for the assisted electrospinning of chitosan are not well known. Therefore, structural characterization of the fiber products was carried out to further understand the structure and composition of these composite materials, as described in further detail below.

3.1. SEM Results

SEM images of Chi:HPCD fiber are shown in Figure 1. We observed a mixture of products that possess nodule-shaped elements and fiber-like morphology in both Figure 1A,B that show the resulting materials for both ratios (2:20 and 2:50) with a heterogeneous composition. It was noted that the Chi:HPCD 2:50 fiber contained less nodule-shape elements and had a large fiber diameter when compared with the Chi:HPCD 2:20 fiber. These general observations coincide with those reported in the previous study by Burns et al. [31].

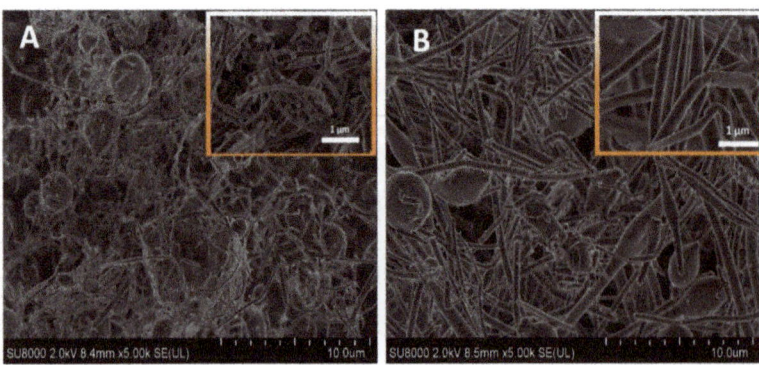

Figure 1. Scanning Electron Microscopy (SEM) images of chitosan (chi)/hydroxypropyl-β-cyclodextrin (Chi:HPCD) fibers. (**A**) Chi:HPCD 2:20 and (**B**) Chi:HPCD 2:50.

3.2. Determination of HPCD Content in the As-Spun Fibers

To assess the composition of the Chi:HPCD electrospun fibers, the HPCD content in the *as-spun* fibers was examined using ^{1}H NMR spectroscopy. The composition of HPCD (wt. %) in the *as-spun* fibers was calculated based on the integration of a –CH$_3$ group of HPCD (δ = 0.99 ppm) relative to the internal standard THF (δ = 1.60 ppm). The ^{1}H NMR spectra are shown in the accompanying Supplementary Materials. To verify the accuracy of the method, a quantitative determination of a known amount of HPCD was evaluated in a test sampling trial. The results are summarized in Table 1, where the measured content (wt. %) of HPCD in the *as-spun* fiber sample was drastically lower than that of the theoretical value (the composition according to mass ratios in the prepared solution). The Chi:HPCD 2:20 fiber had ~50% difference and the Chi:HPCD 2:50 fiber had ~20% difference, as compared with the respective theoretical values based on wt. % content. This suggests that solvent effects may contribute to the weight of the *as-spun* fiber, which yields a reduced composition (wt. %) of HPCD. This difference also infers that there was more solvent residue in the 2:20 fiber than the 2:50 fiber.

Table 1. Determination of hydroxypropyl-β-cyclodextrin (HPCD) content in the *as-spun* Chi:HPCD fiber using ^{1}H NMR spectroscopy.

Material	Determined HPCD Content	Theoretical Value *
HPCD	100% (4.47 mg) [1]	100 ± 6.9% (4.80 mg) [1]
Chi:HPCD 2:20 Fiber	~40%	~91%
Chi:HPCD 2:50 Fiber	~75%	~96%

* Theoretical value was calculated based on mass ratios between chitosan and HPCD in the prepared solution, assuming that all the solvent was evaporated from fibers. [1] Values in parentheses refer to the actual sample weight used for the qNMR calibration (*cf.* Section 2.4).

3.3. FT-IR Results of As-Spun Fibers

The Chi:HPCD fibers were further characterized by FT-IR to identify the composition of the *as-spun* fiber. The FT-IR spectra of *as-spun* fibers are presented in Figure 2 without normalization. FT-IR spectrum of pristine chitosan was shown in Supplementary Materials (Figure S1A). The IR band at 1786 cm^{-1} relates to a free –COOH group in TFA [35–38]. The intensity of this band for Chi:HPCD 2:20 was higher than the value estimated for Chi:HPCD 2:50. This further indicates that there is higher TFA content in the 2:20 fiber than the 2:50 fiber system, which agrees with the results in Table 1. The –COO$^-$ band at 1679 cm^{-1} corresponded to that of the trifluoroacetate ion [35]; whereas the band at 1526 cm^{-1} was assigned to the protonated amine (–NH$_3^+$) group of chitosan [34,39,40]. These signatures indicate that electrostatic interactions are likely to occur between the –COO$^-$ group from TFA with the cation sites (–NH$_3^+$) of chitosan within the fiber composite. The band at 1221 cm^{-1} and the shoulder at 1183 cm^{-1} was assigned to the vibrational signature of C–F for TFA [36–38]. The two spectral signatures disappear in the 2:50 fiber, but were evident for the 2:20 fiber. The spectral variation between samples is the result of interference of free TFA in the 2:20 fiber, as it contains more TFA over the 2:50 fiber, according to the qNMR results in Section 3.2 and the IR band intensity results of the free –COOH group (1786 cm^{-1}) of TFA [35–38]. Hence, the vanishing of two signatures at 1221 and 1183 cm^{-1} in the 2:50 fiber can infer that the –CF$_3$ group of TFA is bound by the annular region of the HPCD host to form a stable complex. An analogous inclusion mode for β-cyclodextrin-guest complexes that were formed between volatile organics with a trifluoromethyl group have been reported by Wilson and Verrall [32] for a range of β-CD/halothane systems (*cf.* Scheme 2 in [32]).

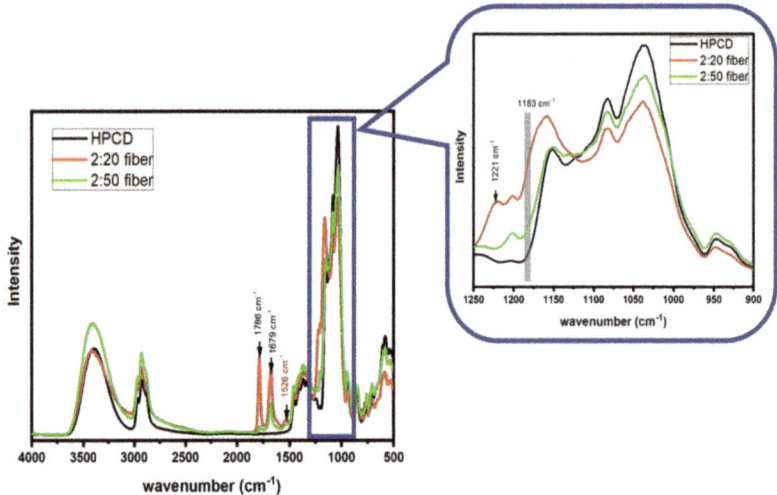

Figure 2. FT-IR spectra of HPCD and *as-spun* Chi:HPCD fiber (2:20 and 2:50, respectively) without normalization. The expanded region of the inset spectra is shown between 1250 and 900 cm^{-1} of the original spectra.

3.4. TGA and DSC Results of Chi:HPCD Fiber

The *as-spun* Chi:HPCD 2:50 fiber was characterized by TGA and DSC, where the 2:20 *as-spun* fiber was not measured because it contained higher free TFA, which was done in order to avoid potential damage to the instrument (DTG plot of chitosan was given in Supplementary Materials, Figure S1B). Differential thermogravimetric (DTG) plots of HPCD and Chi:HPCD 2:50 are shown in Figure 3A, where a thermal event for HPCD at ~350 °C showed no apparent temperature shift, as compared with that of the fiber prepared at the 2:50 ratio. The 2:50 fiber displayed a new thermal event at ~260 °C that corresponds to the decomposition of a trifluoroacetate salt [41]. This indicates that the major fraction of TFA in 2:50 fiber is in the form of a trifluoroacetate salt ($CF_3COO^-/-NH_3^+$), where the $-NH_3^+$ groups are the protonated glucosamine sites of chitosan. This trend is consistent with the FT-IR results reported herein (*cf.* Figure 2). According to the DSC results in Figure 3, evidence of mixing occurs between chitosan and HPCD. However, because of the greater mass content of HPCD in the binary system (chitosan + HPCD), the thermal analysis (TGA and DSC) results for the Chi:HPCD fiber may be obscured, precluding further detailed compositional analysis of the fiber material.

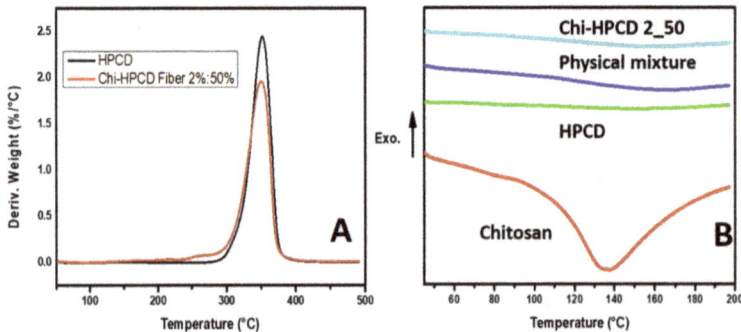

Figure 3. Thermal analysis results of *as-spun* Chi:HPCD fiber. (**A**) DTG plots of HPCD and Chi:HPCD 2:50 fiber; and (**B**) DSC profiles of chitosan, HPCD, physical mixture, and *as-spun* Chi:HPCD 2:50.

3.5. Raman Results

The Raman spectra of HPCD, pristine chitosan, *as-spun* Chi:HPCD fibers and Chi:HPCD fibers after 3 days and 3 months are demonstrated in Figure 4. The band at 730 cm^{-1} and 1788 cm^{-1} was assigned to the deformation of –COO$^-$ and the stretching of carbonyl group in trifluoroacetate anion, respectively [42]. The relative intensity of this band to other spectral signatures revealed a gradual decrease between the *as-spun* fiber (Figure 4C) to an aged fiber after 3 days (Figure 4E), then to a further aged fiber after 3 months (Figure 4F). This demonstrates that TFA content in the fiber decreases slowly over time. A noticeable fact is that signatures related to the –CF$_3$ group (1143, 1202, 601 and 521 cm^{-1}) [42] were not clearly observed in the Raman spectra, in contrast to the –COO$^-$ group of TFA, which suggests that the –CF$_3$ group may form an inclusion complex with HPCD. This conclusion is supported by the FT-IR results, as evidenced by the disappearance of bands at 1221 and 1183 cm^{-1} in Figure 2. Expansions of the original spectra in the range of 770 and 990 cm^{-1} are similarly shown in Figure 4. The Raman signatures over this spectral range relate to the variation of the C–O–C torsional angle, since it is sensitive to its local conformation [34,43]. For HPCD, the band at 850 and 925 cm^{-1} was related to ring breathing of HPCD, while the band at 948 cm^{-1} related to the skeletal vibration for the α-1,4 linkage of HPCD [44]. A comparison was made among precursors (ie, HPCD and chitosan) and different Chi:HPCD fiber types. New bands were observed at 827, 864, 884 and 918 cm^{-1}. By comparison, the band at 950 cm^{-1}, when compared with the spectrum of HPCD in Figure 4A, displayed minimum change. Among the new spectral bands observed, none were associated with the signatures of the CF$_3$COO$^-$ anion when compared with the results reported by Robinson and Taylor's Raman spectral study of the trifluoroacetate ion [42]. These bands showed maximum relative intensity for the *as-spun* Chi:HPCD 2:50 fiber spectrum (Figure 4C) and minimum relative intensity in the spectrum of Chi:HPCD 2:50 fiber after 3 months (Figure 4F). Therefore, new spectral bands at 827, 864, 884 and 918 cm^{-1} may be related to conformational changes of the glucopyranose unit of HPCD that result from the interfacial host–guest complex [32] formed between the –CF$_3$ group of TFA and HPCD. The attenuation of the relative intensity of these spectral bands may be ascribed to attenuated formation of such complexes between TFA and HPCD, as the trifluoroacetate ion content of the fiber decreased. Similar Raman spectral changes have been reported elsewhere [45,46]. Furthermore, the spectral band shapes centered at 850 and 925 cm^{-1} for the Chi:HPCD 2:50 fiber after 3 months in the spectrum (Figure 4F) was similar to the HPCD precursor (Figure 4A), but revealed broader Raman spectral features. By comparison, the band at 948 cm^{-1} showed little difference between HPCD (Figure 4A) and the Chi:HPCD fibers (Figure 4C–F). This effect may suggest that while the –CF$_3$ group of TFA was bound to the interfacial region of HPCD, there was limited host–guest interaction with the α-1,4 linkage domain of HPCD. As well as this, the compositional difference between chitosan and HPCD precluded a diagnostic spectral analysis to assess the details of the host–guest interaction between chitosan and HPCD due to the presence of excess HPCD in the system.

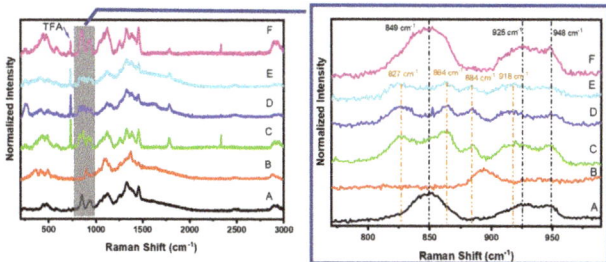

Figure 4. Raman spectra of precursors and different Chi:HPCD fibers: (**A**) HPCD, (**B**) pristine chitosan, (**C**) *as-spun* Chi:HPCD 2:50 fiber, (**D**) *as-spun* Chi:HPCD 2:20 fiber, (**E**) Chi:HPCD 2:50 fiber after 3 days and (**F**) Chi:HPCD 2:50 fiber after 3 months. The expanded region (**Right**) cover the region between 770 and 990 cm^{-1}.

3.6. Raman Imaging with Dye Probe (Rhodamine 6G)

To assess the interaction between chitosan and HPCD, the TFA content was minimized for the Chi:HPCD 2:50 fiber by exposing the fibers to the open atmosphere under adequate ventilation at ambient conditions. This setup favored the volatilization of excess TFA that may influence the sensitivity of the fiber characterization by reducing contributions of high spectral intensity that arise from TFA. The combined use of Raman microimaging along with a suitable dye probe afforded a spectral method that highlighted the chitosan fiber domains that enable the spectral characterization of the regions of interest. Rhodamine 6G was chosen as the dye probe for this study because of the reported favorable binding of chitosan [47] relative to HPCD with the Rhodamine 6G dye [48,49]. Benzene was used as the solvent to avoid the dissolution of fiber components and to maintain the integrity of the electrospun fiber system as it is a poor solvent for such polysaccharide systems (Chi:HPCD). The characteristic Raman signature at 610 and 850 cm^{-1} corresponded to the C–C–C ring *in-plane* bending for Rhodamine 6G [50,51] and the respective ring breathing band for the HPCD glucopyranose unit [44] that was used to construct the 2-D Raman images.

Raman microimaging results in the presence of the Rhodamine 6G dye probe are shown in Figure 5. The Raman spectral image of the HPCD area generated using the Raman signal at 850 cm^{-1} for the fiber is shown in Figure 5A. From this 2-D Raman image, fiber materials with a heterogeneous morphology of beads and fibers concur with the SEM results (Figure 1). The 2-D Raman spectral imaging of the chitosan fraction was generated using the Raman band at 610 cm^{-1}, as shown in Figure 5B. It appears that chitosan adopts a *bundle-type* structure in various sample loci. Upon comparing Figure 5A,B, the spectral region for the chitosan domains were not congruently matched with the spectral region of HPCD and may indicate that the composition of the electrospun fiber was heterogeneous in nature. The occurrence of such heterogeneities likely occurs during the electrospinning process. Despite the efforts that were made to remove TFA from the fiber, the persistence of Raman signatures of TFA were noted at ~730 cm^{-1} but with a reduced spectral intensity.

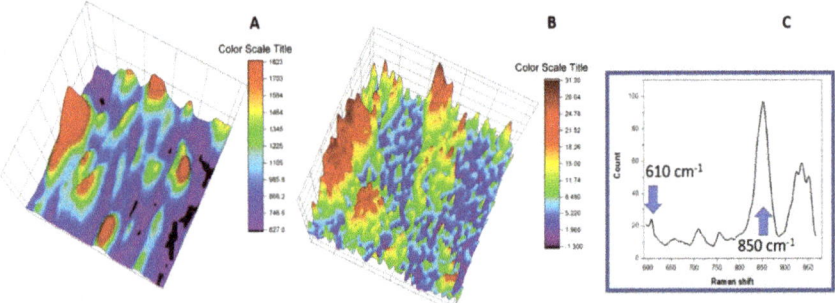

Figure 5. Raman imaging results of dried Chi:HPCD 2:50 fiber after soaking with Rhodamine 6G in a benzene solution, where a sample Raman spectrum centered at 790 cm^{-1} was given under static conditions (λ_{ex} = 785 nm). The Raman spectral image of the HPCD area (**A**) was constructed by peak integration at 850 cm^{-1}. The Raman spectral image of chitosan in area (**B**) was constructed by dividing the spectral intensity for Rhodamine 6G (610 cm^{-1}) against that for HPCD (850 cm^{-1}) for each respective 1-D spectrum, where a 1-D Raman spectrum is shown in panel C as an example.

Figure 5A,B was combined into one additive 2-D image, as shown in Figure 6 to emphasize different structural domains of the fiber: a chitosan rich area, a chitosan/HPCD mixed area and a HPCD rich area. In the spectrum for the chitosan rich domain (Figure 6A), a Raman signature related to the ring breathing mode of chitosan was noted at 895 cm^{-1}. This band had no apparent spectral change when compared against the Raman spectrum of pristine chitosan (Figure 4B). Moreover, the shape of the spectral band at ~850 and ~930 cm^{-1} for all samples (Figure 6A–C) were nearly identical. This may further suggest that HPCD has no direct intermolecular interactions with the chitosan polysaccharide chain of the fiber.

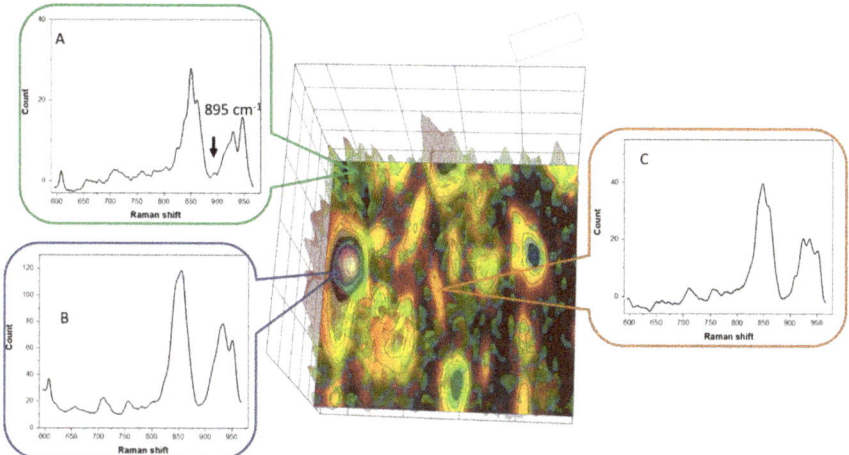

Figure 6. A combined Raman image from the spectral data for the chitosan region (Figure 5A) and the HPCD region (Figure 5B). Sample spectrum was shown for different highlighted spectral regions: (**A**) a chitosan rich domain, (**B**) a chitosan/HPCD mixed domain and (**C**) an HPCD rich domain.

3.7. Composition of Chi:HPCD Electrospun Fiber and Its Component Interactions

An illustration of compositional change of the Chi:HPCD electrospun fiber over time is shown in Scheme 2. According to the FT-IR and Raman spectral results, TFA existed in the Chi:HPCD electrospun fiber as a third component that underwent slow release from the fiber over time that led to composition and/or structural changes of the fiber. Electrostatic interactions were likely to occur between the $-COO^-$ group from the TFA and NH_3^+ group from chitosan under such conditions. A host–guest complex was inferred to occur between the $-CF_3$ group in TFA and HPCD, as described above. This was firstly supported by the disappearance of the C–F band at 1221 and 1183 cm^{-1} according to FT-IR spectra. Secondly, this was confirmed by reduction of the relative intensity of $-CF_3$ signatures in the Raman spectra. Thirdly, further verification was judged by the Raman spectral modifications of the ring breathing region of HPCD for the Chi:HPCD fiber. Thus, the majority of remaining TFA in the fiber may have served as a "connector unit" between chitosan and HPCD, where the $-COO^-$ group of TFA was associated with the NH_3^+ group of chitosan via ion–ion interactions, while the $-CF_3$ group of TFA formed an interfacial complex with HPCD. To afford this type of host–guest complex, the wider annular region that contained the secondary hydroxyl groups of HPCD was implicated, since the narrow annular region contained C6-hydroxyl and C6-hydroxypropyl substituents that may have resulted in steric effects. Upon ion–ion binding of TFA with the charged amine sites ($-NH_3^+$) of chitosan, repulsive electrostatic forces between chitosan polymer chains became attenuated. As well as this, the presence of bound TFA onto chitosan with subsequent binding by HPCD may have afforded additional charge screening that led to improved electrospinning performance of the system, as illustrated in Scheme 2. Based on the 2-D Raman microimaging results, compositional heterogeneity of Chi:HPCD electrospun fiber was depicted, where no evidence of a direct or well-defined host–guest interaction between chitosan and HPCD for the fiber system was supported by the Raman spectral results. In turn, this was consistent with the likely formation of a "facial complex", as reported for the case of a β-CD/halothane complex with a reported 1:1 binding constant ca. 10^2 M^{-1} in aqueous media [32].

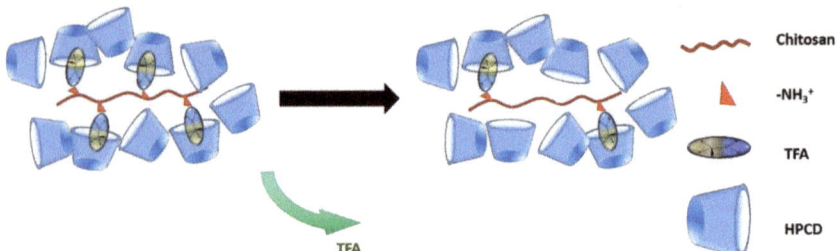

Scheme 2. An illustrative view of the compositional change of a Chi:HPCD electrospun fiber over time, where the green arrow shows incremental temporal loss of trifluoroacetic acid (TFA).

4. Conclusions

Chi:HPCD 2:20 and 2:50 fiber were produced via electrospinning of a mixture of HPCD and chitosan using TFA as a solvent. The composition of Chi:HPCD fibers was determined using thermal analysis and complementary spectral methods. TFA was found to be a constituent in the Chi:HPCD fiber assembly. The heterogeneous morphology and composition of this electrospun fiber was revealed using SEM and Raman imaging, along with a dye-based probe method [34]. Interactions among the components were also characterized using complementary methods, such as IR/Raman spectroscopy. TFA appears to have multifunctional properties as a solvent during the electrospinning process, but also protonates chitosan and stabilizes host–guest interactions with HPCD. Hence, the presence of lateral binding sites along the chitosan backbone (*cf.* Scheme 2) due to electrostatically bound TFA are found to play a crucial role in the effective dispersion of chitosan and the reduction of repulsive forces during electrospinning between chitosan polymer chains that favor fiber formation. There was no clear evidence of direct interactions between the glucosamine moiety of chitosan and HPCD in the solid-state according to Raman imaging, along with the dye probe method. The combined complementary results herein account for the fiber formation process as the role of weak host–guest supramolecular interactions that arise from the formation of a facial complex, which have been independently reported [32]. In turn, we envisage that such controlled-release supramolecular disassembly will contribute to the development of "smart coatings" that may utilize diverse types of chitosan polyelectrolyte complexes.

Supplementary Materials: The following are available online at http://www.mdpi.com/2079-6439/7/5/48/s1, Figure S1: FT-IR spectrum (**A**) and a DTG plot (**B**) of pristine chitosan; Figure S2: ^{1}H NMR spectra of pure HPCD (**A**) and Chi:HPCD 2:20 fiber (**B**) prepared in 1% (*w/w*) tetrahydrofuran (THF)/DMSO-d_6 solution for HPCD content determination; Figure S3: Solid-state ^{13}C CP-TOSS NMR spectra of HPCD (**A**), Chitosan (**B**) and Chi:HPCD (**C**) 2:50 fiber (from top to bottom, respectively).

Author Contributions: Conceptualization, C.X. and L.D.W.; methodology, C.X.; software, C.X.; validation, C.X. and L.D.W.; formal analysis, C.X.; investigation, C.X.; resources, L.D.W.; data curation, C.X.; writing—original draft preparation, C.X.; writing—review and editing, C.X. and L.D.W.; visualization, C.X.; supervision, L.D.W.; project administration, L.D.W.; funding acquisition, L.D.W.

Funding: This research was funded by the Government of Canada through the Natural Sciences and Engineering Research Council (NSERC), Discovery Grant Number: RGPIN 2016-06197.

Acknowledgments: The Natural Sciences and Engineering Research Council (Discovery Grant Number: RGPIN 2016-06197) and the University of Saskatchewan are gratefully acknowledged for support of this research.

Conflicts of Interest: The authors declare no conflict of interest.

References

1. Lin, H.-Y.; Yeh, C.-T. Alginate-crosslinked chitosan scaffolds as pentoxifylline delivery carriers. *J. Mater. Sci. Med.* **2010**, *21*, 1611–1620. [CrossRef] [PubMed]
2. Bhattarai, N.; Edmondson, D.; Veiseh, O.; Matsen, F.A.; Zhang, M.Q. Electrospun chitosan-based nanofibers and their cellular compatibility. *Biomaterials* **2005**, *26*, 6176–6184. [CrossRef] [PubMed]

3. Jayakumar, R.; Prabaharan, M.; Nair, S.V.; Tamura, H. Novel chitin and chitosan nanofibers in biomedical applications. *Biotechnol. Adv.* **2010**, *28*, 142–150. [CrossRef] [PubMed]
4. Min, B.M.; Lee, S.W.; Lim, J.N.; You, Y.; Lee, T.S.; Kang, P.H.; Park, W.H. Chitin and chitosan nanofibers: electrospinning of chitin and deacetylation of chitin nanofibers. *Polymer* **2004**, *45*, 7137–7142. [CrossRef]
5. Kim, S.-K. *Chitin, Chitosan, Oligosaccharides and Their Derivatives: Biological Activities and Applications*; 2011; ISBN 9781439816035. Available online: https://www.vitalsource.com/products/chitin-chitosan-oligosaccharides-and-their-se-kwon-kim-v9781439816042 (accessed on 1 April 2019).
6. Koide, S.S. Chitin-chitosan: Properties, benefits and risks. *Nutr. Res.* **1998**, *18*, 1091–1101. [CrossRef]
7. Elsabee, M.Z.; Naguib, H.F.; Morsi, R.E. Chitosan based nanofibers, review. *Mater. Sci. Eng. C Mater. Biol. Appl.* **2012**, *32*, 1711–1726.
8. Kumar, M.; Muzzarelli, R.A.A.; Muzzarelli, C.; Sashiwa, H.; Domb, A.J. Chitosan chemistry and pharmaceutical perspectives. *Chem. Rev.* **2004**, *104*, 6017–6084. [CrossRef] [PubMed]
9. Kumar, M.N.V.R. A review of chitin and chitosan applications. *React. Funct. Polym.* **2000**, *46*, 1–27.
10. Dash, M.; Chiellini, F.; Ottenbrite, R.M.; Chiellini, E. Chitosan-A versatile semi-synthetic polymer in biomedical applications. *Prog. Polym. Sci.* **2011**, *36*, 981–1014. [CrossRef]
11. Crini, G.; Badot, P.-M. Application of chitosan, a natural aminopolysaccharide, for dye removal from aqueous solutions by adsorption processes using batch studies: A review of recent literature. *Prog. Polym. Sci.* **2008**, *33*, 399–447.
12. Naseri-Nosar, M.; Ziora, Z.M. Wound dressings from naturally-occurring polymers: A review on homopolysaccharide-based composites. *Carbohydr. Polym.* **2018**, *189*, 379–398. [CrossRef] [PubMed]
13. Verma, D.; Katti, K.S.; Katti, D.R. Polyelectrolyte-complex nanostructured fibrous scaffolds for tissue engineering. *Mater. Sci. Eng. C* **2009**, *29*, 2079–2084. [CrossRef]
14. Li, L.; Hsieh, Y.L. Chitosan bicomponent nanofibers and nanoporous fibers. *Carbohydr. Res.* **2006**, *341*, 374–381. [CrossRef]
15. Muzzarelli, R.A.A.; El Mehtedi, M.; Mattioli-Belmonte, M. Emerging Biomedical Applications of Nano-Chitins and Nano-Chitosans Obtained via Advanced Eco-Friendly Technologies from Marine Resources. *Mar. Drugs* **2014**, *12*, 5468–5502. [CrossRef] [PubMed]
16. Rahimi, M.; Emamgholi, A.; Tabaei, S.J.S.; Khodadoust, M.; Taghipour, H.; Jafari, A. Perspectives of chitosan nanofiber/film scaffolds with bone marrow stromal cells in tissue engineering and wound dressing. *Nanomed. J.* **2019**, *6*, 27–34.
17. Aranday-Garcia, R.; Saimoto, H.; Shirai, K.; Ifuku, S. Chitin biological extraction from shrimp wastes and its fibrillation for elastic nanofiber sheets preparation. *Carbohydr. Polym.* **2019**, *213*, 112–120. [CrossRef] [PubMed]
18. Jamshidifard, S.; Koushkbaghi, S.; Hosseini, S.; Rezaei, S.; Karamipour, A.; Rad, A.J.; Irani, M. Incorporation of UiO-66-NH2 MOF into the PAN/chitosan nanofibers for adsorption and membrane filtration of Pb(II), Cd(II) and Cr(VI) ions from aqueous solutions. *J. Hazard. Mater.* **2019**, *368*, 10–20. [CrossRef] [PubMed]
19. Li, L.; Li, Y.; Cao, L.; Yang, C. Enhanced chromium (VI) adsorption using nanosized chitosan fibers tailored by electrospinning. *Carbohydr. Polym.* **2015**, *125*, 206–213. [CrossRef]
20. Habiba, U.; Siddique, T.A.; Talebian, S.; Lee, J.J.L.; Salleh, A.; Ang, B.C.; Afifi, A.M. Effect of deacetylation on property of electrospun chitosan/PVA nanofibrous membrane and removal of methyl orange, Fe(III) and Cr(VI) ions. *Carbohydr. Polym.* **2017**, *177*, 32–39. [CrossRef]
21. Yihan, W.; Minato, W. Nanofiber Fabrication Techniques and Its Applicability to Chitosan. *Prog. Chem.* **2014**, *26*, 1821–1831.
22. Megelski, S.; Stephens, J.S.; Chase, D.B.; Rabolt, J.F. Micro- and nanostructured surface morphology on electrospun polymer fibers. *Macromolecules* **2002**, *35*, 8456–8466. [CrossRef]
23. Casasola, R.; Thomas, N.L.; Trybala, A.; Georgiadou, S. Electrospun poly lactic acid (PLA) fibres: Effect of different solvent systems on fibre morphology and diameter. *Polymer* **2014**, *55*, 4728–4737. [CrossRef]
24. Khil, M.S.; Cha, D.I.; Kim, H.Y.; Kim, I.S.; Bhattarai, N. Electrospun nanofibrous polyurethane membrane as wound dressing. *J. Biomed. Mater. Res. Part B Appl. Biomater.* **2003**, *67B*, 675–679. [CrossRef] [PubMed]
25. Zhao, X.; Chen, S.; Lin, Z.; Du, C. Reactive electrospinning of composite nanofibers of carboxymethyl chitosan cross-linked by alginate dialdehyde with the aid of polyethylene oxide. *Carbohydr. Polym.* **2016**, *148*, 98–106. [CrossRef] [PubMed]

26. Chang, J.-J.; Lee, Y.-H.; Wu, M.-H.; Yang, M.-C.; Chien, C.-T. Preparation of electrospun alginate fibers with chitosan sheath. *Carbohydr. Polym.* **2012**, *87*, 2357–2361. [CrossRef]
27. Ma, G.; Liu, Y.; Peng, C.; Fang, D.; He, B.; Nie, J. Paclitaxel loaded electrospun porous nanofibers as mat potential application for chemotherapy against prostate cancer. *Carbohydr. Polym.* **2011**, *86*, 505–512. [CrossRef]
28. Haider, S.; Al-Zeghayer, Y.; Ali, F.A.A.; Haider, A.; Mahmood, A.; Al-Masry, W.A.; Imran, M.; Aijaz, M.O. Highly aligned narrow diameter chitosan electrospun nanofibers. *J. Polym. Res.* **2013**, *20*, 105. [CrossRef]
29. Wu, C.; Su, H.; Tang, S.; Bumgardner, J.D. The stabilization of electrospun chitosan nanofibers by reversible acylation. *Cellulose* **2014**, *21*, 2549–2556. [CrossRef]
30. Jia, Y.-T.; Gong, J.; Gu, X.-H.; Kim, H.-Y.; Dong, J.; Shen, X.-Y. Fabrication and characterization of poly (vinyl alcohol)/chitosan blend nanofibers produced by electrospinning method. *Carbohydr. Polym.* **2007**, *67*, 403–409. [CrossRef]
31. Burns, N.A.; Burroughs, M.C.; Gracz, H.; Pritchard, C.Q.; Brozena, A.H.; Willoughby, J.; Khan, S.A. Cyclodextrin facilitated electrospun chitosan nanofibers. *RSC Adv.* **2015**, *5*, 7131–7137. [CrossRef]
32. Wilson, L.D.; Verrall, R.E. A volumetric and NMR study of cyclodextrin-inhalation anesthetic complexes in aqueous solutions. *Can. J. Chem.* **2015**, *93*, 815–821. [CrossRef]
33. Dehabadi, L.; Wilson, L.D. Nuclear Magnetic Resonance Investigation of the Fractionation of Water-Ethanol Mixtures with Cellulose and Its Cross-Linked Biopolymer Forms. *Energy Fuels* **2015**, *29*, 6512–6521. [CrossRef]
34. Xue, C.; Wilson, L.D. A structural study of self-assembled chitosan-based sponge materials. *Carbohydr. Polym.* **2019**, *206*, 685–693. [CrossRef] [PubMed]
35. Skrepleva, I.Y.; Voloshenko, G.I.; Librovich, N.B.; Maiorov, V.D.; Vishnetskaya, M.V.; Mel'nikov, M.Y. Infrared spectroscopic studies on interactions in trifluoroacetic acid-sulfur dioxide systems. *Moscow Univ. Chem. Bull.* **2011**, *66*, 232–234. [CrossRef]
36. Valenti, L.E.; Paci, M.B.; De Pauli, C.P.; Giacomelli, C.E. Infrared study of trifluoroacetic acid unpurified synthetic peptides in aqueous solution: Trifluoroacetic acid removal and band assignment. *Anal. Biochem.* **2011**, *410*, 118–123. [CrossRef] [PubMed]
37. Fuson, N.; Josien, M.L.; Jones, E.A.; Lawson, J.R. Infrared and raman spectroscopy studies of light and heavy trifluoroacetic acids. *J. Chem. Phys.* **1952**, *20*, 1627–1634. [CrossRef]
38. Redington, R.L.; Lin, K.C. Infrared spectra of trifluoroacetic acid and trifluoroacetic anhydride. *Spectrochim. Acta Part A Mol. Spectrosc.* **1971**, *27*, 2445–2460. [CrossRef]
39. Sun, W.; Chen, G.; Wang, F.; Qin, Y.; Wang, Z.; Nie, J.; Ma, G. Polyelectrolyte-complex multilayer membrane with gradient porous structure based on natural polymers for wound care. *Carbohydr. Polym.* **2018**, *181*, 183–190. [CrossRef]
40. Kasaai, M. The Use of Various Types of NMR and IR Spectroscopy for Structural Characterization of Chitin and Chitosan. In *Chitin, Chitosan, Oligosaccharides and Their Derivatives*; CRC Press: Boca Raton, FL, USA, 2010; pp. 149–170, ISBN 9781439816042.
41. Mosiadz, M.; Juda, K.L.; Hopkins, S.C.; Soloducho, J.; Glowacki, B.A. An in-depth in situ IR study of the thermal decomposition of yttrium trifluoroacetate hydrate. *J. Therm. Anal. Calorim.* **2012**, *107*, 681–691. [CrossRef]
42. Robinson, R.E.; Taylor, R.C. Raman spectrum and vibrational assignments for the trifluoroacetate ion. *Spectrochim. Acta* **1962**, *18*, 1093–1095. [CrossRef]
43. Frech, R.; Chintapalli, S.; Bruce, P.G.; Vincent, C.A. Crystalline and Amorphous Phases in the Poly(ethylene oxide)–LiCF$_3$SO$_3$ System. *Macromolecules* **1999**, *32*, 808–813. [CrossRef]
44. Egyed, O. Spectroscopic studies on β-cyclodextrin. *Anal. Chim. Acta* **1990**, *240*, 225–227. [CrossRef]
45. Li, W.; Lu, B.; Chen, F.; Yang, F.; Wang, Z. Host-guest complex of cypermethrin with β-cyclodextrin: A spectroscopy and theoretical investigation. *J. Mol. Struct.* **2011**, *990*, 244–252. [CrossRef]
46. De Oliveira, V.E.; Almeida, E.W.C.; Castro, H.V.; Edwards, H.G.M.; Dos Santos, H.F.; De Oliveira, L.F.C. Carotenoids and β-cyclodextrin inclusion complexes: Raman spectroscopy and theoretical investigation. *J. Phys. Chem. A* **2011**, *115*, 8511–8519. [CrossRef] [PubMed]
47. Vanamudan, A.; Pamidimukkala, P. Chitosan, nanoclay and chitosan-nanoclay composite as adsorbents for Rhodamine-6G and the resulting optical properties. *Int. J. Biol. Macromol.* **2015**, *74*, 127–135. [CrossRef]

48. Serra-Gómez, R.; Tardajos, G.; González-Benito, J.; González-Gaitano, G. Rhodamine solid complexes as fluorescence probes to monitor the dispersion of cyclodextrins in polymeric nanocomposites. *Dyes Pigment.* **2012**, *94*, 427–436. [CrossRef]
49. Bakkialakshmi, S.; Menaka, T. A study of the interaction between rhodamine 6g and hydroxy propyl β-cyclodextrin by steady state fluorescence. *Spectrochim. Acta Part A Mol. Biomol. Spectrosc.* **2011**, *81*, 8–13. [CrossRef] [PubMed]
50. Dieringer, J.A.; Wustholz, K.L.; Masiello, D.J.; Camden, J.P.; Kleinman, S.L.; Schatz, G.C.; Van Duyne, R.P. Surface-enhanced Raman excitation spectroscopy of a single rhodamine 6G molecule. *J. Am. Chem. Soc.* **2009**, *131*, 849–854. [CrossRef] [PubMed]
51. Black, P.N.; He, X.N.; Zhou, Y.S.; Mahjouri-Samani, M.; Mitchell, M.; Gao, Y.; Allen, J.; Lu, Y.F.; Xiong, W.; Jiang, L. Surface-enhanced Raman spectroscopy using gold-coated horizontally aligned carbon nanotubes. *Nanotechnology* **2012**, *23*, 205702.

© 2019 by the authors. Licensee MDPI, Basel, Switzerland. This article is an open access article distributed under the terms and conditions of the Creative Commons Attribution (CC BY) license (http://creativecommons.org/licenses/by/4.0/).

Article

Development of Oxygen-Plasma-Surface-Treated UHMWPE Fabric Coated with a Mixture of SiC/Polyurethane for Protection against Puncture and Needle Threats

Dariush Firouzi [1], Chan Y. Ching [1], Syed N. Rizvi [2] and P. Ravi Selvaganapathy [1,*]

1. Department of Mechanical Engineering, McMaster University, Hamilton, ON L8S 4L7, Canada; dariush.firouzi@gmail.com (D.F.); chingcy@mcmaster.ca (C.Y.C.)
2. RONCO, 70 Planchet Road, Concord, ON L4K 2C7, Canada; rizvi.n.syed@gmail.com
* Correspondence: selvaga@mcmaster.ca; Tel.: +1-(905)-525-9140 (ext. 27435)

Received: 28 March 2019; Accepted: 13 May 2019; Published: 20 May 2019

Abstract: Although considerable research has been directed at developing materials for ballistic protection, considerably less has been conducted to address non-firearm threats. Even fewer studies have examined the incorporation of particle-laden elastomers with textiles for spike, knife, and needle protection. We report on a new composite consisting of ultra-high-molecular-weight polyethylene (UHMWPE) fabric impregnated with nanoparticle-loaded elastomer, specifically designed for spike- and needle-resistant garments. Failure analysis and parametric studies of particle-loading and layer-count were conducted using a mixture of SiC and polyurethane at 0, 30, and 50 wt.%. The maximum penetration resistance force of a single-layer of uncoated fabric increased up to 218–229% due to nanoparticle loading. Multiple-layer stacks of coated fabric show up to 57% and 346% improvement in spike puncture and hypodermic needle resistance, respectively, and yet were more flexible and 21–55% thinner than a multiple-layer stack of neat fabric (of comparable areal density). We show that oxygen-plasma-treatment of UHMWPE is critical to enable effective coating.

Keywords: coating; UHMWPE; nanoparticle-laden elastomer; oxygen-plasma treatment; penetration resistance

1. Introduction

In many countries around the world, non-firearm weapons have been associated with the dominant fraction of violent crimes. For instance, in Canada, about 19% of all violent crimes committed in 2016 involved the use of non-firearm weapons while only 3% were attributed to the use of firearms [1]. Due to the increasing number of assaults committed using knives and sharpened instruments, puncture and stab resistance has become increasingly important as a safety feature for body armor [2]. Resistant materials capable of providing protection against sharp objects are also of interest in a number of different occupations, such as waste processing. There is an ever-present threat of stab and knife attacks for law enforcement and military personnel [3], as well as incidences of accidental injuries due to sharp debris, broken glass, or razor wire [4] to health care and waste management workers. The main parameters which affect the penetration resistance of textiles are the areal density of fabrics, yarn linear density and mechanical properties of fibers, number of filaments per yarn, fabric weave architecture, and the number of fabric plies. There are other parameters that could affect the resistance of a multilayer textile structure against various types of threats or penetrators, such as the boundary condition of the fabric, geometry, type, mass, and the velocity of penetrator, and inter-yarn and fabric-projectile frictional force [5–8].

Typical textile-based resistant materials comprise of high-strength fibers (HSFs), such as aramid, S-glass, ultra-high molecular weight polyethylene (UHMWPE), polypyridobisimidazole (PIPD), and polyphenylene benzobisoxazole (PBO). HSFs generally possess high tensile and compressive strength, high energy absorption, and low density [9,10]. UHMWPE fibers and fabrics are used widely in the protection industry for military body armors, helmets, and personal protective equipment (PPE), such as cut-resistant gloves [11,12]. Spectra® (Honeywell International Inc., Morristown, NJ, USA) and Dyneema® (Royal DSM) are the two main commercially produced UHMWPE fibers that have the lowest density of all HSFs currently being used for armor applications [9]. Additionally, UHMWPE possesses excellent chemical resistance and also high resistance against physical degradation [13], which makes this material a promising substitute for Kevlar® (DuPont) for body protection. Despite the many advantages of using UHMWPE fibers and fabrics for armor applications, there are considerably fewer research studies of this polymer as compared to Kevlar® (DuPont) for body armor application. Here, we exclusively summarize previous research works related to the use of UHMWPE fabric for improved penetration resistance against non-ballistic threats (such as stab, puncture, and needle penetration) and discuss related mechanisms.

Coating fabric with polymeric materials without particle addition is one simple method to improve its penetration resistance. For example, puncture resistant UHMWPE fabrics have been developed using Nylon 6,6 and Nylon 6,12 coatings. The normalized spike puncture resistance of a multiple-layer stack of UHMWPE fabric was improved up to 62% from nylon coating due to the mechanical-interlocking mechanism of the nylon solution upon washing and drying [2,14].

The impregnation of textiles with shear thickening fluids (STFs) is a relatively new approach to improve the penetration resistance of HSFs [15–22]. Interestingly, most of these studies use Kevlar® fabric and only a few have focused on the use of considerably lower-cost and lighter UHMWPE fabrics. In one particular study, it was shown that the effectiveness of STF-impregnation in improving the penetration resistance of UHMWPE fabric is dependent on fabric structure. For instance, STF-impregnation was a detrimental treatment for penetration resistance (against a spherical-head impactor at 4.5 m/s) of 400-denier UHMWPE fabrics (various thread counts except 25 × 25). On the contrary, STF-impregnation of 1350-denier UHMWPE provides up to 59% higher normalized penetration resistance force [15]. The effectiveness of STF-impregnation on UHMWPE fabric against spike and knife penetration also depends on the choice of dispersing medium used in the fabrication of STF. For instance, UHMWPE fabric impregnated with polyethylene glycol (PEG400)-based STF material even has a slightly lower stab resistance force to spike threat compared to neat fabrics [16]. Apart from the varying performances reported, one potential limitation of STF-treatment technology is the propensity of fluid-like STF materials to leakage [17], as they could be easily washed away when exposed to water and moisture [18]. In general, the mechanism of improvement of penetration resistance of STF-treated textiles is a combination of impact energy absorption due to shear thickening phenomenon and also increased inter-yarn friction [15,19].

More recently, a different technology using particle-laden elastomeric mixtures with textiles has been examined for enhanced penetration resistance. In one study, it was shown that high density polyethylene (HDPE) fabric impregnated with a mixture of heavily-loaded silica and SiC nano-sized particles (65 wt.% and 80 wt.%) in polydimethylsiloxane (PDMS) mixture provides up to 190% increase in normalized penetration resistance force to 21G hypodermic needles when it was compared to the neat fabric. The improvement in penetration resistance of treated fabrics was related to the elastic-jamming phenomenon at a very high concentration of particles [3]. Despite their promise, such nano-particle elastomer composites are difficult to incorporate into UHMWPE fabrics and to the best of authors' knowledge, studies on spike puncture resistance of particle-elastomer impregnated UHMWPE fabrics have not been reported elsewhere.

Here, we show that oxygen plasma-treatment is an effective method for impregnating the nanoparticle elastomer material into UHMWPE fabrics. Using this method, we demonstrate for the first time new textile composites that have superior spike and needle puncture resistance and study the influence on the composition of these composites on the resistance conferred. Various samples were developed for penetration tests using UHMWPE fabric impregnated with different mixtures of SiC particles in a polyurethane (PU)-based elastomer. Our developed formulation and coating method has a good potential to be applied to protective garments or gloves.

2. Penetration Tests and Characterization

2.1. Materials

The UHMWPE plain woven fabric (Spectra® 900) with an areal density of 230 g/m^2, 1200 denier, 1333 dtex, and thread-count of 21 × 21 yarns per inch was supplied by Barrday Inc., Charlotte, NC, USA. The silicon carbide (SiC) particles from Panadyne Inc. Montgomeryville, PA, USA, had a nominal diameter of 0.5 µm and specific surface area of 4–8 m^2/g. A polyurethane-based elastomer (Brush-On® 40 kit, Smooth-On, Inc., Macungie, PA, USA) with 1000% elongation at break and density of 1170 kg/m^3 was purchased from Smooth-On, Inc., Macungie, PA, USA. The kit was comprised of A and B components to be mixed in a 1:1 ratio by weight. An organic solvent (mineral spirit) was purchased from Recochem Inc., Milton, ON, Canada.

2.2. Fabric Preparation-Plasma Treatment

UHMWPE fabric was cut to 12 cm × 18 cm and then put in a capacitively-coupled systems plasma chamber (PE-25, Plasma Etch, Inc., Carson City, NV, USA) with a radio frequency (RF) source at 50 kHz and vacuumed at around 18×10^{-2} Torr. Oxygen was then fed into the chamber (50 cc/min) for 60 s during gas stabilization phase. The oxygen feed was then closed before the RF power was applied for 5 min with power of 300 W. The plasma treatment was found to increase the adhesion of UHMWPE to coating materials.

2.3. Coating Preparation

Organic solvent (mineral spirit) was added to the pre-polymer base part of Brush-On® 40 kit, so the weight ratio of solvent to part A was 3.5 and was mixed thoroughly for 2 h. SiC particles were then added to the mixture to 30 wt.% and 50 wt.% of the final mixture (i.e., the total weight of particles and parts A and B of the elastomer) and mixed for 3 h. Paste-like part B of the elastomer kit was then added to the well-dispersed mixture and stirred thoroughly for 30 min. At this point, the mixture had a low viscosity, which made an immersion (dip) coating process feasible without additional equipment. The plasma-treated fabric samples were soaked (dipped) in the mixture for 15 s and hang-dried for five hr. Using this technique, the material pickup of fabric samples from a same coating mixture is fairly equivalent, although the samples were not squeezed after the dip process. One general advantage of the dip coating technique is that fibers are not stressed or distorted during such a process [23]. Figure S1 in Supplementary Information shows a schematic of the fabrication process and plasma treatment setup.

2.4. SpikePuncture and NeedlePenetration Tests

Spike puncture penetration tests were performed according to the European standard for protective gloves against mechanical risks (EN 388:2016standard), which is also suggested by the American National Standard Institute/International Safety Equipment Association (ANSI/ISEA 105-2016 standard). Using a SHIMADZU tensile tester with 500 N load cell, a steel stylus-like spike (Figure 1b) with a diameter of 4.5 mm (nose diameter of 1 mm) was pushed through the fabric sample at a traverse speed of 100 mm/min. The supporting fixture for the fabric was made in a similar manner to the EN 388 standard, with an open internal diameter of 20 mm (Figure 1a). The fabric sample was held (sandwiched) firmly between the supporting fixture and another plate with the same open internal diameter, using four clamps at the directions of warp and weft fibers. Each test was repeated at least ten times and the maximum resistance force was recorded. In a similar fashion, multiple-layer stacks of fabric were tested.

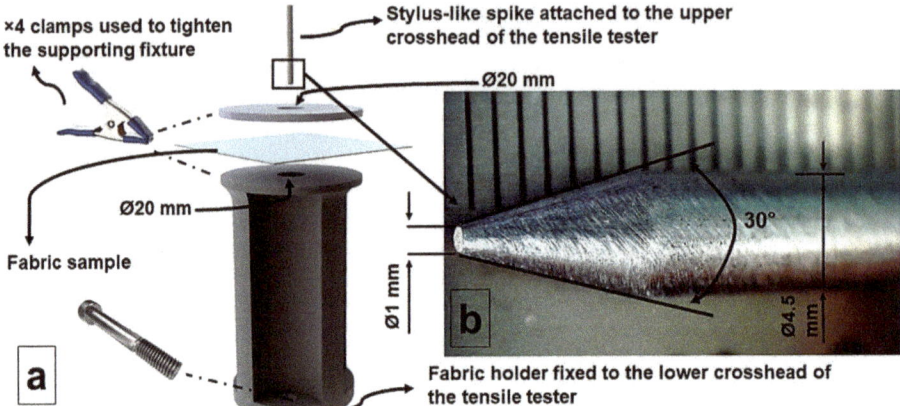

Figure 1. (a) Schematic of penetration test setup. (b) Stylus-like spike, according to EN 388:2016 standard (millimeter-scale).

Target samples were then categorized into different levels based on their maximum resistance against spike puncture (Table S1 in Supplementary Information). Hypodermic needle penetration tests were performed similarly to a previous study [3] and in accordance with the American Society for Testing and Materials (ASTM F2878–10 standards). Briefly, the fabric sample was held firmly between the supporting fixture and another plate with the same open internal diameter of 20 mm. The fabric target was then punctured by the needles at a speed of 500 mm/min. A fresh 25G 1" (BD 305125) hypodermic needle was used for each test and repeated at least ten times and the maximum resistance force was recorded. Samples were then categorized into different levels based on their maximum resistance against hypodermic needle according to ANSI/ISEA 105-2016 (Table S1 in Supplementary Information).

2.5. Flexibility Test and Coefficient of Friction Determination

Flexibility tests were performed similar to our previous work [3], based on the principles of ASTM D1388. The objective of this test was to compare the rigidity of samples before and after coating. A 5 cm × 10 cm piece of each sample was cut and encapsulated inside a thin polyethylene bag and a 20 g weight was attached to its leading edge at mid-point. With 6.2 cm length of the sample held firmly on the edge of table, its bending angle (BA) is then measured from captured images. A higher bending angle from this test means higher flexibility of the sample. Static coefficient of friction of neat fabrics (before and after plasma treatment) along the weft direction were calculated according to ASTM D1894-14, using a SHIMADZU tensile tester with 500 N load cell (Method-C of apparatus for assembly).

2.6. Image Analysis and Contact Angle Measurement

The images were taken using a handheld digital microscope (Celestron Acquisition, LLC, Torrance, CA, USA). Scanning electron microscopy (SEM) analysis was done using a VEGA-LSU TESCAN VP. Optical contact angle measurements were done using a DataPhysics OCA-35 instrument to analyze wettability of the fabric surface before and after plasma treatment. A drop of Milli-Q® (MilliporeSigma, Burlington, VT, USA) ultrapure water (dosing rate of 2 µL/s) was applied using the instrument's electronic syringe unit on the surface of the sample and the corresponding contact angle was recorded.

3. Results and Discussion

An initial spike test was conducted on as-received and coated plasma-treated UHMWPE fabric samples to show the effectiveness of the coating process. The images of the samples before and after puncture under a maximum pre-set load of 100 N are shown in Figure 2. The top row shows the spike penetration of a 3-layer stack of neat fabric and the bottom row shows the same penetration test on a 2-layer stack of oxygen-plasma-treated coated fabric (with 50 wt.% of SiC in PU mixture) of a comparable areal density. It can be seen that the 3-layer neat fabric sample was easily penetrated, and fibers were pushed aside at the impact area. This is a typical "windowing mechanism" in textiles under the impact of a spike or similar sharp pointed objects [2,3]. On the contrary, the 2-layer stack of coated fabric sample was not penetrated and only a small dent was visible on the surface.

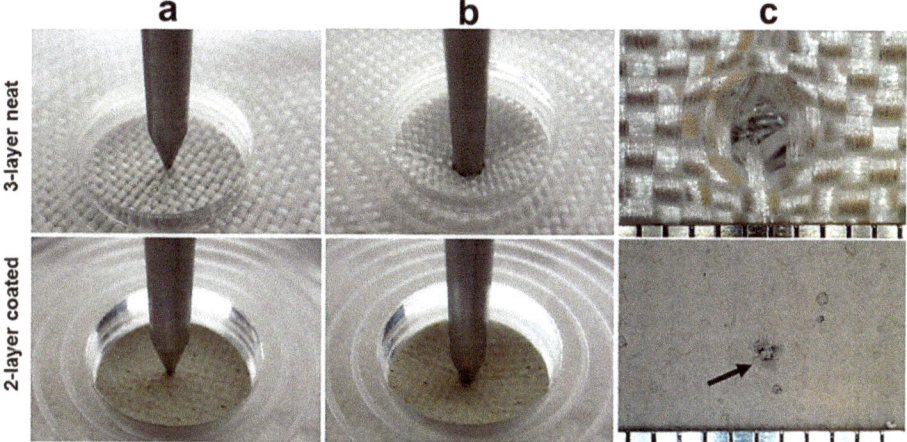

Figure 2. Images of the fabric samples undergoing the spike penetration test under a maximum pre-set load of 100 N: (**a**) start of penetration; (**b**) end of penetration; (**c**) top close view of the impact area. The top row is a 3-layer stack of neat fabric, while the bottom row is 2-layer stack of coated fabric with 50 wt.% SiC in polyurethane (PU). The arrow shows the point of contact on the coated sample (millimeter-scale).

In order to understand the results better, the material was microscopically analyzed. The SEM images of untreated fabric with only PU coating are shown in Figure 3a. It is seen from the figure that the coating material is not uniform and poorly bonded to the fabric substrate. Similar observations can be made for the coated fabric with 30 and 50 wt.% SiC in a PU mixture from Figure 3b,c, respectively. On the other hand, plasma-treated fabric with PU coating (Figure 3d) showed improved coating uniformity. Similar improvement in uniformity of the coating material was also observed in particle-laden PU coatings (comparing Figure 3e to Figures 3b and 3f to Figure 3c). The areal density of coated samples increased by about 2–5% when the fabrics were plasma-treated and coated, which indicates a better wettability and higher absorption of the coating material onto fibers. Since plasma treatment has been used in the past to open reactive groups on textile surfaces to functionalize them [23], this behavior is expected. UHMWPE fiber with low surface energy and a smooth surface is inherently inert and not a good host to bond with other materials [2,13]. Through plasma treatment, polarizable groups can be introduced that will enhance wettability and also bonding to other matrix materials. For instance, plasma treatment has been used to improve adhesion between UHMWPE and HDPE matrices, which resulted in higher interlaminar shear strength (ILSS), tensile strength, and impact strength of the fabricated composites [24]. Similarly, plasma treatment of UHMWPE fibers also improves interfacial bonding with epoxy resin, resulting in higher tensile strength but considerably lower elongation [13]. In general, oxidative RF-plasma treatment could effectively improve wettability and surface characteristics of polyethylene by creating oxygen-based functionalities [25]. More specifically, the surface modification of UHMWPE through oxygen plasma-treatment is likely due to the formation of carbon–oxygen functional groups (such as C–O, C=O, O=C–O) and cross-linking of UHMWPE molecules, which contribute to the increased surface free energy and hydrophilic transformation of the oxygen-plasma-treated polymer [26,27]. The water contact angle of the as-received UHMWPE fabric was 120.7 ± 3.6°, which can be considered as a hydrophobic material. However, when a droplet of water was dropped on the surface of the UHMWPE fabric, which was treated with plasma for 1/6min, 1/2min, 1min, 2min, 3min, 4min, and 5 min, the water instantly spread over the surface of the fabric, indicating complete wetting of the plasma-treated fabric. The change of contact angle confirms the remarkable change in surface properties from an inert material to one that has polarizable groups and open dangling bonds, which could be beneficial for wettability and adhesion of coatings to the UHMWPE fabric (Figure 3a–f).

Figure 3. (a–f) Top view before-puncture test scanning electron microscope (SEM) images of plasma-treated and not-treated ultra-high molecular weight polyethylene (UHMWPE) fabrics coated with different mixtures of SiC in PU. The scale bar is 500 μm. (g,h) Plasma-treated and not-treated UHMWPE fibers. The scale bar is 10 μm.

The SEM images of UHMWPE fibers after plasma treatment (Figure 3h) reveal the presence of micro-pits onto the surface, while the as-received fiber appears perfectly smooth (Figure 3g). The presence of such micro-pits onto the surface of plasma-treated fibers could improve adhesion of UHMWPE fibers through a mechanical interlocking mechanism [28]. This observation is in agreement with the results for the coefficient of friction. The static coefficient of friction of UHMWPE fabric increased from 0.20 ± 0.08 to 0.38 ± 0.12 when it was plasma-treated for 5 min due to its higher roughness, which was due to the presence of micro-pits.

3.1. Puncture Test Results

A single-layer neat plasma-treated UHMWPE fabric and single-layer samples of plasma-treated UHMWPE fabric coated with 0 wt.%, 30 wt.%, or 50 wt.% concentrations of SiC were tested to study the influence of the particle loading on puncture resistance against a spike threat. The typical force-displacement curves of 1-layer plasma-treated neat and coated plasma-treated fabrics against spike penetration are shown in Figure 4a. The resistance force increases when the tip of the penetrator comes in contact with each of the samples. A first load peak occurs at about 3–6 mm of penetration distance in all three coated samples, which matches the length of the conical tip of the spike shown in Figure 1b. At the first peak, an audible burst sound was heard, which is correlated to the breakage of adjacent fibers as the tip of the spike pierced through the samples [2]. The average resistance force of the coated plasma-treated fabric at the first peak load is 28.3 ± 4.3 N, 55.3 ± 8.2 N, and 67.3 ± 6.8 N when the concentration of SiC is 0wt.%, 30wt.%, and 50 wt.%, respectively. This means that the resistance force of primary fibers in contact with the spike at the impact area increases when more SiC is added to the mixture of PU coating.

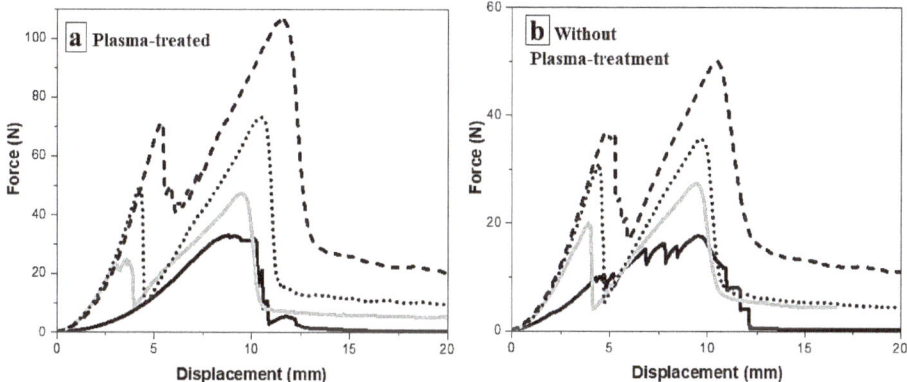

Figure 4. Force-versus-displacement curves from the spike penetration test on single-layer UHMWPE fabric samples: —, neat; — (gray color), coated with PU; •••, coated with 30 wt.% SiC in PU; —, coated with 50 wt.% SiC in PU.

After the first peak, there is a sudden drop in the force in all three coated samples followed by an increase till the force reaches a second peak. The first peak and the sudden drop can be correlated to the point when the conical tip of the spike completely passed through the sample. The increase in the penetration force after the first peak can be attributed to the resistance from distant fibers near the impact zone. The average resistance force at the second peak-load (maximum resistance force) is 45.7 ± 4.5 N, 67.3 ± 8.3 N, and 105.7 ± 9.3 N when the concentration of SiC was 0 wt.%, 30 wt.%, and 50 wt.%, respectively.

The load drops after the second peak as the conical tip of the spike completely penetrates through the fabric and adjacent fibers provide minimal further resistance. The plasma-treated neat fabric shows only a single peak of 33.2 ± 8.7 N. In this case, unlike the coated fabric samples, the distant fibers do not exert a force to the spike penetrator when the failed fibers are pushed aside. This behavior can be attributed to the lower fiber-to-fiber and fiber-to-impactor frictional forces when the UHMWPE fabric is not coated.

Additionally, Figure 4b shows the force-displacement curves from spike penetration tests of neat and coated fabric samples without prior oxygen-plasma treatment. Comparing these results with those in Figure 4a, plasma treatment increases the spike penetration resistance force of one-layer coated and uncoated fabric samples. The penetration resistance force of one-layer neat and coated fabric with 0 wt.%, 30 wt.%, and 50 wt.% SiC in PU decreased by about 35%, 47, 51%, and 54%, respectively, when the fabric was not plasma-treated.

Visual images of both the top and rear of the penetrated fabric (with or without prior plasma treatment) were taken and are shown in Figure 5 along with SEM images. In the case of the fabric coated with only the elastomer (only PU), the top view did not show any meaningful difference as compared to the rear view (Figure 5a,d), and therefore it was chosen to be represented. From the visual images (Figure 5a–f), it was observed that all the three plasma-treated samples show a good coverage of the nanoparticle/elastomer material. On the contrary, the coating material is not uniform when the fabric is not plasma-treated prior to the coating process. A closer view of the damage zone is shown in Figure 5g–l. It can be clearly seen from Figure 5g that the PU coating forms a shell on top of the untreated fabric, while in Figure 5j the PU is well integrated with the fibers of the fabric. A similar contrast is also observed between the untreated (Figure 5h,i) and the treated (Figure 5k,l) fabrics as the particle loading in the elastomer mixture is increased to 30% and then 50%. These results demonstrate that the plasma treatment plays a critical role in the penetration of the nanoparticle-loaded PU onto the individual filaments, as opposed to the untreated case, where it forms a thin shell with poor bonding to the fibers. The integration of the nanoparticles into the depth of the fabric and between filaments is crucial in providing higher resistance to puncture. Furthermore, the coating material was found to be increasingly more damaged and peeled-off when the loading of nanoparticles was increased from 0 wt.% to 50 wt.% of the PU mixture onto fabric without prior plasma treatment. From Figure 5j–l, the effectiveness of plasma-treatment to improve adhesion and bonding strength of coating material to UHMWPE fabric is significant at all the loading concentrations of SiC nanoparticles in PU mixture. Overall, the extent of the damage is lesser for the plasma-treated fabric rather than the untreated one.

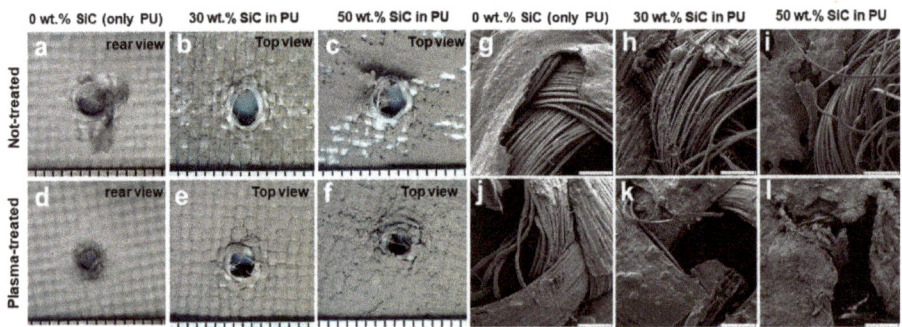

Figure 5. (a–f) Top and rear views of plasma-treated and not-treated UHMWPE fabrics coated with different mixtures of SiC in PU after penetration. (g–l) SEM images (top view) of plasma-treated and not-treated UHMWPE fabrics coated with different mixtures of SiC in PU. The scale bar is 500 μm.

In one study, Kevlar 129® and UHMWPE fabric were coated with natural rubber (NR) and tested against a cylindrical shape penetrator with a hemispherical head at 6 m/s. It was reported that the absorbed impact energy absorption of a single-layer rubber-coated fabric was even lower than that of a neat fabric due to a lower yarn-to-penetrator frictional force [29]. On the contrary, here we showed the improvement in maximum penetration resistance of a single-layer UHMWPE fabric at a high concentration of SiC nanoparticles. One explanation is that the fibers are more coupled and yarn-to-penetrator frictional force increases with an increase in the concentration of hard nano-sized SiC particles, and thus more force is exerted to penetrate coated fabrics. In addition, the hard nanoparticles that could be loosened during penetration could also increase frictional force in a manner similar to the three-body abrasion, enhancing the overall resistance.

Puncture Test Results of Multiple-Layer Stack of Fabric Samples

Multiple layers of neat and coated fabric samples were tested to investigate the effectiveness of the coating material based on areal density, flexibility, and thickness. For this purpose, several layers of fabric samples were stacked together and tested against spike penetration. The histogram of maximum resistance force, areal density, and flexibility of plasma-treated neat and coated UHMWPE fabric samples of different stacked layers is presented in Figure 6. Among all stacks of neat fabric, only the 7-layer stack of neat fabric could provide Level5 protection. The 7-layer stack of neat fabric has comparable areal density with the following stacked samples: (i) 5-layer coated fabric with PU; (ii) 5-layer coated fabric with 30 wt.% SiC in PU; and (iii) 2-layer coated fabric with 50 wt.% SiC in PU.

The maximum resistance force of these stacked samples was found to be about 20%, 57%, and 15% more, respectively, as compared with the resistance of a 7-layer neat fabric. Interestingly, the 5-layer coated fabric with 30 wt.% of SiC in PU showed about 36% higher penetration force compared with the 2-layer coated fabric with 50 wt.% SiC in PU (of comparable areal density), although one layer of the coated fabric with 50 wt.% SiC showed the highest maximum penetration resistance among all the samples. This result indicates that the number of stacked layers also play a major role in increasing the penetration resistance of target samples, in addition to the loading of particles in the PU mixture.

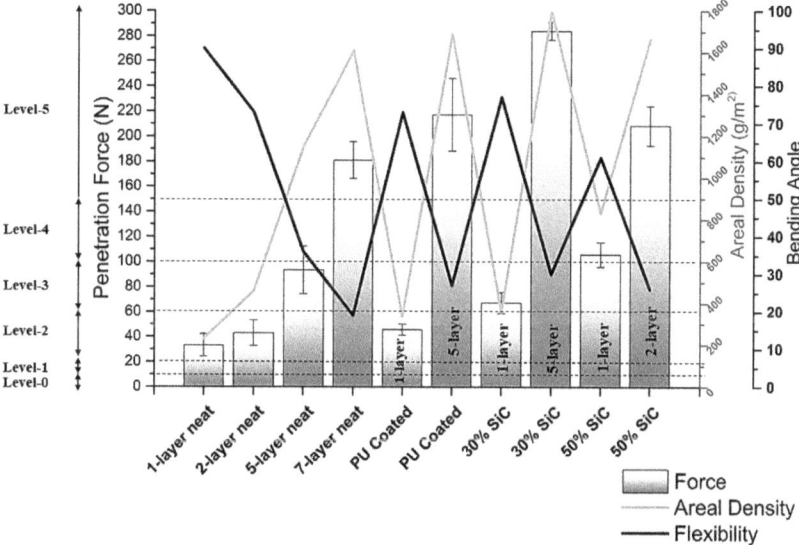

Figure 6. Measured peak resistance force, bending angle, and areal density of plasma-treated neat and coated UHMWPE fabrics of various layers undergoing spike penetration tests.

Also, from Table 1, the average thickness of a 7-layer stack of neat fabric reduced by about 21–55% compared to 2-layers and 5-layers of coated samples of comparable areal density. From Table 1, 1-layer neat fabric has the lowest rigidity, while 1-layer coated fabric with 50 wt.% SiC has the highest. Comparing the flexibility of multi-layer stacks of fabric samples with comparable areal density, it was found that the 7-layer stack of neat fabric was less flexible than 5-layer stack of coated samples with 0 wt.% and 30 wt.% of SiC, and even the 2-layer stack of coated fabric with 50 wt.% SiC in PU mixture.

Table 1. Areal density, thickness, and bending angle of plasma-treated neat and coated UHMWEPE fabric samples.

No. of Stacked Fabric Layers	Particle Loading in PU Mixture (wt.%)	Areal Density (g/m^2)	Thickness (mm)	Bending Angle ($\theta \pm 2°$)
1	Neat (no coating)	230 ± 11	0.5	90
	0 (only PU)	338 + 16	0.51 ± 0.01	73
	30	360 ± 22	0.55 ± 0.01	77
	50	832 ± 59	0.78 ± 0.09	61
7	Neat (no coating)	~1610	3.5	19
5	0 (only PU)	~1690	~2.55	27
	30	~1800	~2.75	30
2	50	~1664	~1.56	26

Apart from areal density, flexibility and thinness are equally important factors in the manufacture of protective gloves [30]. Therefore, a new figure of merit was formulated that represents the ratio of percentage increase in penetration resistance to percentage increase in thickness of different multi-layer neat and coated samples over a single layer of neat fabric and can be used to evaluate the performance of the various compositions and combinations that were fabricated. This figure of merit was found to be 0.74, 1.35, 1.68, and 2.49, respectively, for 7-layer neat, 5-layer coated with PU, 5-layer coated with 30 wt.% SiC, and 2-layer coated with 50 wt.% SiC fabrics. These results show that addition of SiC nanoparticles increases the penetration resistance significantly more as compared with any increases in thickness of the overall coated fabric. Hence, the coated sample with 50 wt.% SiC provides the optimal performance among all the samples, which is Level-5 resistance, acceptable thickness, and flexibility, without the consideration of increase in areal density.

Next, normalized penetration force of each sample was calculated based on its areal density (Table S2 in Supplementary Information). From the results, the normalized force of only 1-layer coated fabric with 50 wt.% particles is higher than 1-layer neat fabric (about 30% improvement). Furthermore, 5-layer coated fabric with 0 wt.% and 30 wt.% particles and 2-layer coated fabric with 50 wt.% particles were found to have about 14%, 40%, and 11% (respectively) higher normalized resistance force compared with a 7-layer neat fabric. In contrast, Kevlar-wool and Kevlar-wool-nylon fabrics had a surprising notable decrease in normalized penetration resistance ($N/(g/m^2)$) against a pointed impactor when they were coated with 20 wt.% silica particles (of 242 μm diameter) and a PVC binder [31]. This may be due to the higher concentration (30 wt.% and 50 wt.%) of particles used in our study to fabricate particle-laden elastomeric mixtures. The 0.5 μm SiC particles used here could also provide a better mixing homogeneity in elastomeric mixture compared with a mixture of 242 μm SiO_2 particles. Additionally, our nanocomposite also outperforms another study [32], where STF materials were impregnated into UHMWPE fabric to improve penetration resistance. There, it was reported that an 8-layer stack of STF-treated UHMWPE fabric provides approximately 21% higher normalized penetration resistance force (against a spike) compared to an 8-layer stack of neat UHMWPE fabric, and yet was about 30% thicker than the neat one [32]. Similarly, STF-treated UHMWPE composite panels have been compared with untreated UHMWPE composite panels under a drop tower impact test (using a spike-like impactor). From the results, the maximum resistance force of STF-treated UHMWPE composite panels was even lower than untreated panels and penetration resistance was not improved after STF treatment [20].

3.2. Hypodermic Needlestick Penetration

Single and multiple layers of neat and coated fabric samples of plasma-treated fabric were tested to study the impact of coating (0 wt.%, 30 wt.%, and 50 wt.% concentration of SiC) on penetration resistance against a needle threat. The typical force-displacement curves of 1-layer plasma-treated neat and coated plasma-treated fabric against needle penetration are shown in Figure 7. The measured resistance force increases as the tip of the needle comes in contact with the fabric samples. The maximum force was detected at about 1–2 mm after the tip of the needle penetrates through the coated fabric samples, which matches the sharp-cutting continuous edge of the needle (2 mm). A peak force represents the maximum resistance of textile samples against penetration of hypodermic needles by spreading the fibers away from the tip (windowing mechanism) and cutting the fibers with the sharp conical part (cutting mechanism) [3,30]. The load drops afterward, with minimal resistance against the penetrator. Unlike load-displacement curves shown for the spike test, a second peak was not detected in the needle test. This can be attributed to the lack of resistance of distant fibers near the point of impact due to the small size of the needle compared with the spike penetrator. Maximum resistance force, bending angle, and areal density histogram plots of 1-layer and multiple-layer neat and coated fabric samples are presented in Figure 8.

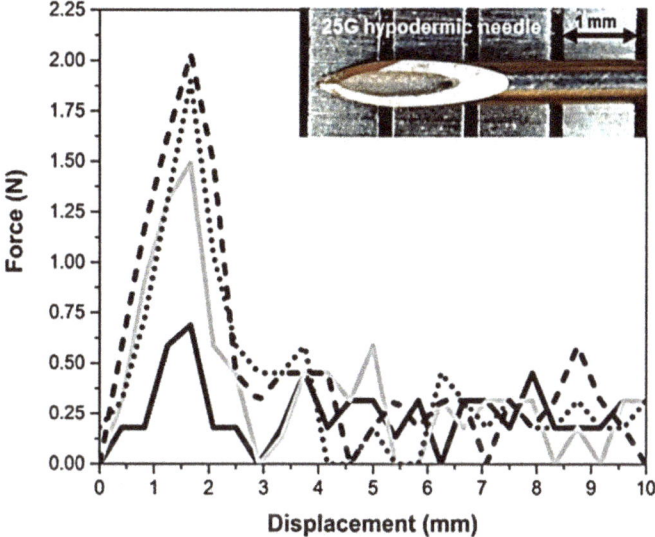

Figure 7. Force-versus-displacement graphs from the 25G needlestick penetration test on single-layer plasma-treated UHMWPE fabric samples: —, neat; —, coated with PU; •••, coated with 30 wt.% SiC in PU; - - -, coated with 50 wt.% SiC in PU.

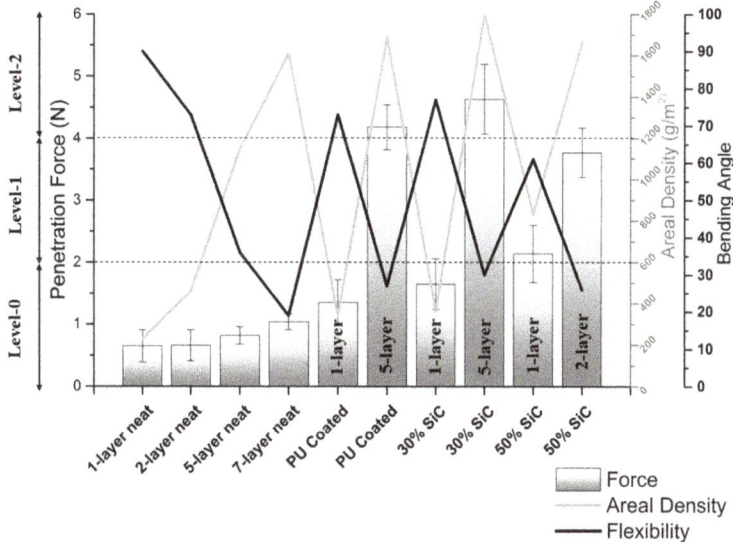

Figure 8. Measured peak resistance force, bending angle, and areal density of plasma-treated neat and coated UHMWPE fabric of various layers undergoing 25G needle penetration tests.

It is observed that the resistance force of 1-layer neat fabric (0.65 ± 0.26 N) does not change significantly compared to 2-layer, 5-layer, and 7-layer neat samples, and even a 7-layer stack of neat fabric is still Level-0 resistance. Coating a 1-layer fabric with 0 wt.%, 30 wt.%, and 50 wt.% SiC in PU mixture increases average penetration resistance by about 108%, 154%, and 229%, respectively.

Furthermore, the normalized penetration resistance force of 1-layer coated fabric with 0 wt.% and 30 wt.% particle was found to be 41% and 62% higher, respectively, than 1-layer neat fabric. Among the coated samples, only 1-layer coated fabric with 50 wt.% SiC is considered as Level-1 protection. However, the normalized force of 1-layer coated fabric with 50 wt.% particle is lower than the neat one (Table S3 in Supplementary Information).

Multiple layers of neat and coated samples were tested to investigate the effectiveness of the coating material based on areal density, flexibility, and thickness. For this purpose, several layers of fabric samples of comparable areal density were stacked together and tested against needlestick penetration. The normalized penetration resistance force of a 7-layer neat fabric (0.64×10^{-3} N/(g/m^2)) increased significantly due to coating, which is about 283, 298, and 251% for 5-layer coated fabric with PU, 5-layer coated fabric with 30 wt.% of SiC, and 2-layer coated fabric with 50 wt.% of SiC, respectively. The highest resistance obtained was at level-2 for the 5-layer coated fabric with PU and also 5-layer coated fabric with 30 wt.% of SiC. These findings show that addition of SiC containing PU coating to UHMWPE does improve needle penetration resistance but only to level-2 (Table S1 in Supplementary Information). In contrast, we have shown previously that particle-laden elastomeric mixtures impregnated into high-count (60 × 60) HDPE can provide a significant increase in penetration resistance (190%) and produce a level-5 resistance level towards needle penetration threats [3] through an elastic jamming mechanism. The differences in the results may largely be related to the fabric used. The lower thread-count of the UHMWPE (21 × 21) here results in the lower penetration force for the neat fabric as compared to the HDPE [3]. The higher baseline resistance enabled the HDPE fabric to achieve level-5 performance when the nanoparticle composite was incorporated. Nevertheless, the looser weave of the UHMWPE fabric allows better incorporation of the nanoparticle/PU mixture, and therefore a higher percentage increase in normalized penetration resistance was observed. Such an interpretation is analogous to a similar study that interprets more effective response of innately infirm UHMWPE fabrics to STF-impregnation [15].

4. Conclusions

We showed the considerable impact of coating UHMWPE fabric with a mixture of SiC and polyurethane (PU) to improve spike and hypodermic penetration resistance. A simple dip coating process to incorporate nanoparticle-elastomer mixtures with up to 50 wt.% particle loading was also demonstrated. Spike puncture tests performed on coated fabric show a significantly higher resistance than neat fabric and the resistance increased with increase in SiC content. The penetration resistance of coated fabric was also higher than the neat fabric, even when different stacks of samples of comparable areal density were compared. Further, 5-layer and 2-layer stacks of coated fabric with a mixture of 30 wt.% and 50 wt.% SiC in PU showed up to 57% higher spike-puncture resistance and better flexibility compared to a 7-layer stack of neat fabric of comparable areal density, with 21–55% less thickness. The higher resistance of coated samples can be attributed to a combination of penetration of the coating into the fabric, better adhesion of coating material with the fibers, and the effect of the loaded nanoparticles in exerting more frictional force to the impactor. We also showed that plasma treatment is a crucial preparation process for UHMWPE fabric to improve the adhesion and integration of coating material to the fabric. Non-uniform coatings and poor penetration and puncture resistances were obtained without plasma treatment. Similarly, the resistance of coated fabrics against a 25G hypodermic needle was superior to the neat fabric (up to 298% increase in normalized penetration resistance due to the treatment). The combination of SiC nanoparticles with PU elastomer provides an interesting solution for increasing the puncture resistance while not affecting the needle penetration significantly. In the future, this property could be used on UHMWPE textiles, which have rather weak resistance against puncture threats, in order to improve their robustness and overall performance against a multitude of threats.

Supplementary Materials: The following are available online at http://www.mdpi.com/2079-6439/7/5/46/s1, Table S1: Classification for puncture resistance for hand protection against a spike and hypodermic 25G needle according to the American National Standard Institute/International Safety Equipment Association (ANSI/ISEA 105-2016) standards. Table S2: Absolute and normalized penetration resistance force of plasma-treated neat and coated UHMWPE fabrics against a spike. Table S3: Absolute and normalized penetration resistance force of plasma-treated neat and coated UHMWPE fabrics against a 25 G hypodermic needle. Figure S1: The schematic of fabrication process and plasma treatment setup.

Author Contributions: D.F. was involved in conceiving the idea, performed experiments, analyzed results, wrote the first draft of the paper; P.R.S. and C.Y.C. were involved in conception of the idea, analyzed results, revised the manuscript; S.N.R. contributed to discussions and revised manuscript.

Funding: This research was funded by the Natural Sciences and Engineering Research Council of Canada (NSERC) under the Discovery and Engage programs. It was also supported by the Canada Research Chairs Program.

Conflicts of Interest: The authors declare that there is no conflict of interest regarding the publication of this article.

References

1. Cotter, A. Firearms and Violent Crime in Canada. Statistics Canada. Available online: https://www150.statcan.gc.ca/n1/pub/85-005-x/2018001/article/54980-eng.htm (accessed on 28 June 2018).
2. Firouzi, D.; Foucher, D.A.; Bougherara, H. Nylon-Coated Ultra High Molecular Weight Polyethylene Fabric for Enhanced Penetration Resistance. *J. Appl. Polym. Sci.* **2014**, *131*, 40350. [CrossRef]
3. Firouzi, D.; Russel, M.K.; Rizvi, S.N.; Ching, C.Y.; Selvaganapathy, P.R. Development of flexible particle-laden elastomeric textiles with improved penetration resistance to hypodermic needles. *Mater. Des.* **2018**, *156*, 419–428. [CrossRef]
4. Decker, M.K.; Halbach, C.J.; Nam, C.H.; Wagner, N.J.; Wetzel, E.D. Stab resistance of shear thickening fluid (STF)-treated fabrics. *Compos. Sci. Technol.* **2007**, *67*, 565–578. [CrossRef]
5. Cheeseman, B.A.; Bogetti, T.A. Ballistic impact into fabric and compliant composite laminates. *Compos. Struct.* **2003**, *61*, 161–173. [CrossRef]
6. Park, J.; Yoon, B.; Paik, J.; Kang, T. Ballistic performance of *p*-aramid fabrics impregnated with shear thickening fluid; Part I—Effect of laminating sequence. *Text. Res. J.* **2012**, *82*, 527–541. [CrossRef]
7. Park, J.; Yoon, B.; Paik, J.; Kang, T. Ballistic performance of *p*-aramid fabrics impregnated with shear thickening fluid; Part II—Effect of fabric count and shot location. *Text. Res. J.* **2012**, *82*, 542–557. [CrossRef]
8. Sanhita, D.; Jagan, S.; Shaw, A.; Pal, A. Determination of inter-yarn friction and its effect on ballistic response of para-aramid woven fabric under low velocity impact. *Compos. Struct.* **2015**, *120*, 129–140.
9. Lane, R.A. High performance fibers for personnel and vehicle armor systems. *Amptiac Q.* **2005**, *9*, 3–9.
10. Langston, T. An analytical model for the ballistic performance of ultra-high molecular weight polyethylene composites. *Compos. Struct.* **2017**, *179*, 245–257. [CrossRef]
11. Zhang, B.; Nian, X.; Jin, F.; Xia, Z.; Fan, H. Failure analyses of flexible ultra-high molecular weight polyethylene (UHMWPE) fiber reinforced anti-blast wall under explosion. *Compos. Struct.* **2018**, *184*, 759–774. [CrossRef]
12. Kurtz, S.M. Chapter 17—Composite UHMWPE Biomaterials and Fibres. In *UHMWPE Biomaterials Handbook*, 2nd ed.; Academic Press: Burlington, MA, USA, 2009; pp. 249–258.
13. Lin, S.P.; Han, J.L.; Yeh, J.T.; Chang, F.C.; Hsieh, K.H. Surface modification and physical properties of various UHMWPE-fiber-reinforced modified epoxy composites. *J. Appl. Polym. Sci.* **2007**, *104*, 655–665. [CrossRef]
14. Firouzi, D.; Youssef, A.; Amer, M.; Srouji, R.; Amleh, A.; Foucher, D.A.; Bougherara, H. A new technique to improve the mechanical and biological performance of ultra high molecular weight polyethylene using a nylon coating. *J. Mech. Behav. Biomed. Mater.* **2014**, *32*, 108–209. [CrossRef]
15. Arora, S.; Majumdar, A.; Butola, B.S. Structure induced effectiveness of shear thickening fluid for modulating impact resistance of UHMWPE fabric. *Compos. Struct.* **2019**, *210*, 41–48. [CrossRef]
16. Li, W.; Xiong, D.; Zhao, X.; Sun, L.; Liu, J. Dynamic stab resistance of ultra-high molecular weight polyethylene fabric impregnated with shear thickening fluid. *Mater. Des.* **2016**, *102*, 162–167. [CrossRef]
17. Wagner, N.; Wetzel, E.D. Advanced Body Armor Utilizing Shear Thickening Fluids. U.S. Patent 7498276 B2, 3 March 2009.
18. Rao, H.M.; Hosur, M.V.; Jeelani, S. Chapter 12—Stab characterization of STF and thermoplastic-impregnated ballistic fabric composites. In *Advanced Fibrous Composite Materials for Ballistic Protection*; Chen, X., Ed.; Woodhead Publishing: Sawston, UK, 2016; pp. 363–387.

19. Kalman, D.P.; Merrill, R.L.; Wagner, N.J.; Wetzel, E.D. Effect of particle hardness on the penetration behavior of fabrics interclated with dry particles and concentrated particle-fluid suspensions. *J. Appl. Polym. Sci.* **2009**, *1*, 2602–2612.
20. Asija, N.; Chouhan, H.; Gebremeskel, S.A.; Bhatnagar, N. Impact response of shear thickening fluid (STF) treated high strength polymer composites—Effect of STF intercalation method. *Proc. Eng.* **2017**, *173*, 655–662. [CrossRef]
21. Asija, N.; Chouhan, H.; Gebremeskel, S.A.; Singh, R.K.; Bhatnagar, N. High strain rate behavior of STF-treated UHMWPE composites. *Int. J. Impact Eng.* **2017**, *110*, 359–364. [CrossRef]
22. Majumdar, A.; Laha, A.; Bhattacharajee, D.; Biswas, I. Tuning the structure of 3D woven aramid fabrics reinforced with shear thickening fluid for developing soft body armour. *Compos. Struct.* **2017**, *178*, 415–425. [CrossRef]
23. Joshi, M.; Butola, B.S. Application technologies for coating, lamination and finishing of technical textiles. In *Advances in Dyeing and Finishing of Technical Textiles*; Woodhead Publishing Ltd.: Cambridge, UK, 2013; pp. 355–411.
24. He, R.; Niu, F.; Chang, Q. The effect of plasma treatment on the mechanical behavior of UHMWPE fiber-reinforced thermoplastic HDPE composite. *Surf. Interf. Anal.* **2017**, *50*, 73–77. [CrossRef]
25. Lehocky, M.; Drnovska, H.; Lapcikova, B.; Barros-Timmons, A.M.; Trindade, T.; Zembala, M.; Lapcik, L., Jr. Plasma surface modification of polyethylene. *Coll. Surf. A Physicochem. Eng. Asp.* **2003**, *222*, 125–131. [CrossRef]
26. Liu, H.; Xie, D.; Qian, L.; Deng, X.; Leng, Y.X.; Huang, N. The mechanical properties of ultrahigh molecular weight polyethylene (UHMWPE) modified by oxygen plasma. *Surf. Coat. Technol.* **2011**, *205*, 2697–2701. [CrossRef]
27. Lee, S.G.; Kang, T.J.; Yoon, T.H. Enhanced interfacial adhesion of ultra-high molecular weight polyethylene (UHMWPE) by oxygen plasma treatment. *J. Adhes. Sci. Technol.* **1998**, *12*, 731–748. [CrossRef]
28. Moon, S.I.; Jang, I. The effect of the oxygen-plasma treatment of UHMWPE fiber on the transverse properties of UHMWPE-fiber/vinylester composites. *Compos. Sci. Technol.* **1999**, *59*, 487–493. [CrossRef]
29. Roy, R.; Laha, A.; Awasthi, N.; Majumdar, A.; Butola, B.S. Multi layered natural rubber coated woven P-aramid and UHMWPE fabric composites for soft body armor application. *Polym. Compos.* **2017**, *39*, 3636–3644. [CrossRef]
30. Houghton, J.M.; Schiffman, B.A.; Kalman, D.P.; Wetzel, E.D.; Wagner, N.J. Hypodermic needle puncture of shear thickening fluid (STF)—treated fabrics. In Proceedings of the SAMPE, Baltimore, MD, USA, 3–7 June 2007; pp. 1–11.
31. Kanesalingam, S.; Nayak, R.; Wang, L.; Padhye, R.; Arnold, L. Stab and puncture resistance of silica-coated Kevlar-wool and Kevlar-wool-nylon fabrics in quasistatic conditions. *Text. Res. J.* **2018**, 1–17. [CrossRef]
32. Sun, L.L.; Xiong, D.S.; Xu, C.Y. Application of shear thickening fluid in ultra high molecular weight polyethylene fabric. *J. Appl. Polym. Sci.* **2013**, *129*, 1922–1928. [CrossRef]

© 2019 by the authors. Licensee MDPI, Basel, Switzerland. This article is an open access article distributed under the terms and conditions of the Creative Commons Attribution (CC BY) license (http://creativecommons.org/licenses/by/4.0/).

Article

One-Step Surface Functionalized Hydrophilic Polypropylene Meshes for Hernia Repair Using Bio-Inspired Polydopamine

Noor Sanbhal [1,2], Xiakeer Saitaer [1,3], Mazhar Peerzada [2], Ali Habboush [4], Fujun Wang [1] and Lu Wang [1,*]

1. Key Laboratory of Textile Science and Technology of Ministry of Education, College of Textiles, Donghua University, 2999 North Renmin Road, Songjiang, Shanghai 201620, China; xaker2@163.com (X.S.); wfj@dhu.edu.cn (F.W.)
2. Department of Textile Engineering, Mehran University of Engineering and Technology, Jamshoro, Sindh 76062, Pakistan; Mazhar.peerzada@faculty.muet.edu.pk; noor.sanbhal@faculty.muet.edu.pk (N.S.)
3. College of Textiles and Fashion, Xingjiang University, 666 Sheng Li Road, Tian Shan, Wulumuqi 830046, China
4. Mechanical Textile Engineering, Aleppo University, Aleppo 15310, Syria; alitex2008@hotmail.com
* Correspondence: wanglu@dhu.edu.cn; Tel./Fax: +86-21-6779-2637

Received: 27 November 2018; Accepted: 8 January 2019; Published: 14 January 2019

Abstract: An ideal hernia mesh is one that absorbs drugs and withstands muscle forces after mesh implantation. Polypropylene (PP) mesh devices have been accepted as a standard material to repair abdominal hernia, but the hydrophobicity of PP fibers makes them unsuitable to carry drugs during the pre-implantation of PP meshes. In this study, for the first time, one-step functionalization of PP mesh surfaces was performed to incorporate bio-inspired polydopamine (PDA) onto PP surfaces. All PP mesh samples were dipped in the same concentration of dopamine solution. The surface functionalization of PP meshes was performed for 24 h at 37 °C and 80 rpm. It was proved by scanning electron microscopic (SEM) images and Fourier Transform Infrared Spectroscopy (FTIR) results that a thin layer of PDA was connected with PP surfaces. Moreover, water contact angle results proved that surface functionalized PP meshes were highly hydrophilic (73.1°) in comparison to untreated PP mesh surfaces (138.5°). Thus, hydrophilic PP meshes with bio-inspired poly-dopamine functionalization could be a good choice for hernia mesh implantation.

Keywords: polypropylene; hernia meshes; surface functionalization; polydopamine

---

1. Introduction

A hernia is a defect in the abdominal wall due to the "protrusion" of an organ [1,2]. However, repair of the hernia is a common practice to reinforce the operated muscles of the hernia to their original place with biomaterial [3]. For this reason, numerous bio-materials have been used to repair hernia [4], for example, gold sutures, silver wires, and nylon prosthetics, but the recurrence of hernia has been a major problem due to the improper selection of material and its design [5]. However, the suture repair of hernia was associated with a higher chance of hernia recurrence compared to mesh implantation [6,7]. Nevertheless, patient pain and surgical site infection is still a major problem of hernia repair [8–10]. Recently, a few synthetic mesh materials, such as polytetrafluoroethylene (PTFE), polyethylene Terephthalate (PET), and polypropylene (PP), have been successfully used to reduce the hernia recurrence rate [11–13] and among them, light weight PP mesh material has been considered an effective material to reduce the hernia recurrence rate [14–16].

However, PP mesh material is hydrophobic in nature and does not absorb drugs during mesh implantation. In the literature, it is reported that standard PP mesh is not suitable for hernia repair due to the reason that pre-operative prophylaxis during mesh implantation has failed [17–19]. Therefore, there is an urgent need to surface functionalize and improve the surface wettability of PP meshes.

Thus, an ideal mesh may be one that is flexible and easy to place during a hernia operation, does not show a prolonged inflammatory response after mesh implantation, does not degrade easily, does not shrink after implantation, is free of adhesion formation or fistulas, and is resistant to infection [20–25]. Faulk et al. coated PP meshes with non-biodegradable hydrogel and reported that coated PP meshes did not demonstrate any sign of infection, foreign body response, and fibrosis [26]. Additionally, Perez-Kohler et al. reported that PP meshes coated with quaternary ammonium compounds successfully resisted mesh infection [27]. Moreover, Bellon et al. stated that partially absorbable PP mesh devices are superior to standard PP meshes in terms of more compliance, being less rigid, being strong, good tissue ingrowth, and being resistant to the infection [28].

The prime objective of the preparation of medical devices is ensuring their biocompatibility [29,30]. Therefore, bio-inspired materials are given more attention, such as dopamine with self-polymerization properties and generating poly-dopamine onto surfaces of any structure and material [31,32].

Dopamine surface functionalization of material is an easy method at around room temperature and pH 8.5. Thus, the surface functionalization process with dopamine may be controlled with the dopamine concentration or process reaction time [33,34]. Therefore, considering the surface properties of chemically inert PP fibers and self-polymerization properties of bio-inspired poly-dopamine, it was our objective to surface functionalize the PP surfaces using poly-dopamine (PDA) without changing the bulk properties of PP mesh fibers.

In this work, PP mesh materials were surface functionalized with PDA at room temperature. PDA was incorporated for 12 and 24 h. The PDA functionalized PP meshes were analyzed using a scanning electron microscope (SEM), Energy Dispersive X-ray Spectroscopy (EDX), Fourier Transform Infrared Spectroscopy (FTIR), Contact angle (sessile drop method), Differential Scanning Calorimeter (DSC), and X-ray diffractometer (XRD). The results confirmed the functionalization of PP meshes with a thin layer of poly-dopamine (PDA). Moreover, the surface wettability of PDA treated PP meshes was dramatically increased in comparison to un-treated PP meshes.

2. Materials and Methods

2.1. Materials

Polypropylene meshes of a light weight (27 g/m^2) were used for surface functionalization with polydopamine (PDA). These medical textile devices were obtained from Nantong Chemical fiber Co. Ltd. China (Nantong, China). Dopamine hydrochloride and tris-(hydroxylmethyl aminomethane) were received from Aladdin Chemicals Ltd Shanghai China.

2.2. Surface Functionalization of PP Meshes Materials with Polydopamine

The dopamine solution (10 mM) was prepared using 0.1 g of dopamine in 50 mL of tris- and maintained at pH 8.5, as described in a recently published paper [33]. During preparation, the liquor ratio of PP mesh materials to dopamine solution was 1:100. PP mesh devices were soaked in a prepared solution of dopamine. The solution was stirred at 80 rpm in a controlled environment at 37 °C for 12–24 h. After the required duration time (12–24 h), PP mesh devices were taken out and hot rinsed (50 °C) with distilled water several times and dried in an oven at 40 °C. All PP mesh devices were functionalized with the same concentration of solution, but different time durations (12 and 24 h).

3. Characterization

3.1. SEM & EDX

PP control and surface functionalized PP meshes were coated using platinum (pt). Platinum coated PP meshes were scanned for surface morphology, using a scanning electron microscope (SEM,

Quanta SEM 250, and FEI™). Moreover, for element analysis, Energy Dispersive (ISIS 300, Oxfordshire, UK) X-ray spectroscopy was used, while an EDX component was attached with SEM.

3.2. FTIR

PP control and dopamine surface functionalized PP meshes were analyzed using Fourier Transform Infrared Spectroscopy (ATR) by Nicolet 6700, Waltham, MA, USA. The samples were scanned in the range of 500–4000 wavenumber cm^{-1}.

3.3. XRD and DSC Analysis

XRD was used to scan PDA functionalized and untreated samples in the range of 2θ (5°–60°). Thus, an X-ray diffractometer made in Tokyo Japan (Rigaku D/MAX 2550/PC) was used at a scanning rate of 0.02°/min. Moreover, a differential scanning calorimeter (Pyris, Perkin Elemer 4000, Grove, IL, USA) was used in the heating range of 25–250 °C to obtain the melting temperature of untreated and PDA functionalized PP meshes.

3.4. Water Contact Angle

PP meshes were of a 0.1 mm fiber diameter and it was difficult to obtain the contact angle of treated and untreated fibers. Therefore, nonwoven fabrics (melt blown, 23 g/m^2) were used to measure the contact angle of PDA treated and untreated fabrics. A dynamic contact angle method using a sessile drop was used to measure the contact angle. Moreover, WCA 20 (software) was employed to calculate the hydrophilic or hydrophobic water contact angle of PDA functionalized and untreated PP fabrics. Each time, three drops (5 µL) were dispensed onto fabric and an average value was calculated.

3.5. Statistical Analysis

The standard deviation and error bars are presented in the figures with symbols (* and —). One way analysis of variance (ANOVA) was employed to analyze the data. The data with *** is less than 0.001 and mentioned as $p < 0.001$, while the data with (**) represents $p < 0.01$ and the data displaying (*) means $p < 0.05$. Thus, the value of p (*) < 0.05 was selected as the confidence interval value.

4. Results and Discussion

4.1. Surface Functionalization of PP Mesh Fibers

PP meshes were surface functionalized with bio-inspired polydopamine (PDA). Thus, PP meshes were soaked in a weak alkaline solution of dopamine and continuously stirred for 12–24 h at 37 °C. However, PDA was expected to coat surfaces of PP mesh by self-polymerization. The process of soaking and PDA structure are shown in Figure 1. According to the literature, PDA can form a thin layer on the surface of any fiber [35]. Thus, we received similar results and PP mesh fibers were successfully surface functionalized with PDA.

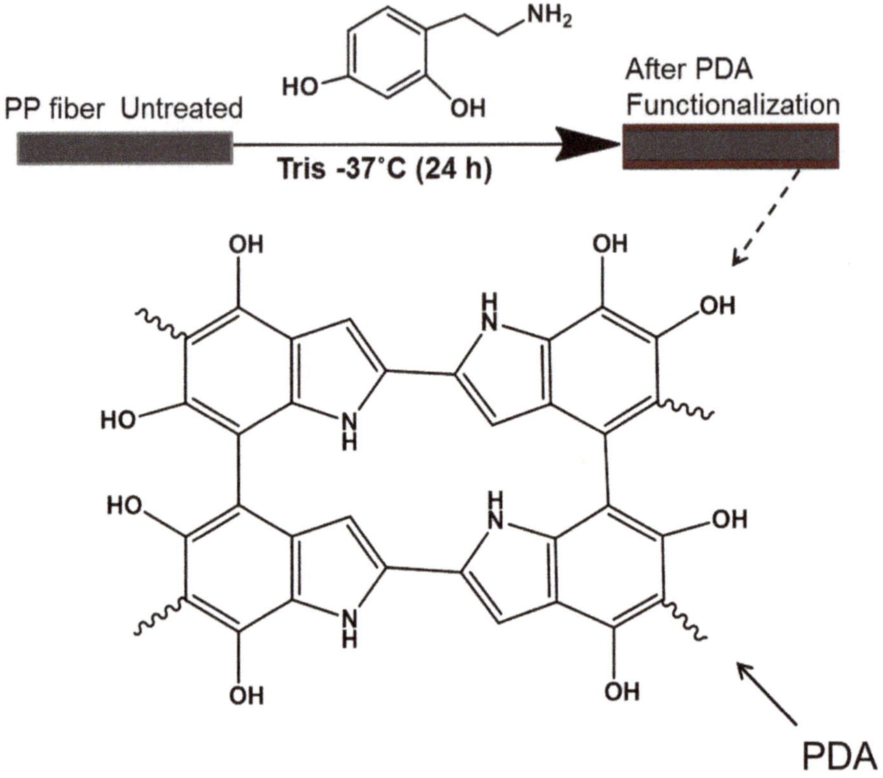

Figure 1. Schematic of PP meshes soaking process in dopamine and PDA structure.

Moreover, PP meshes were soaked in the same concentration of dopamine solution, but for different durations of time. We observed a marginal difference in the surface morphology and coating efficiency between the 12 h soaking and 24 h soaking time. The PP meshes soaked for a 24 h dipping time exhibited a rougher surface with more PDA compared to that of PP meshes dipped for 12 h. However, there was not much difference in the yield. During the 12 h dipping duration, 0.2% of an average weight was increased, while during the 24 h dipping time, 0.35% corresponding weight was increased. It was really hard to record an accurate average weight increase; therefore, we considered standard testing atmosphere conditions (37 °C and 65% humidity) and weighed the samples before and after the surface functionalization of PP mesh fibers.

4.2. Surface Morphology of PP Meshes Fibers Before and After Surface Functionalization

Figure 2 displays SEM images of PDA surface functionalized and untreated PP meshes. It can be seen that the PP control (Figure 2a,b) shows smooth surfaces before functionalization, but after PDA treatment for 12 h, a thin layer of PDA coated the surfaces of PP fibers (Figure 2b,c). PDA treated PP mesh surfaces displayed a dark grey color with a shiny surface, rather than a dull grey color. Moreover, small spheres on the surfaces of treated fibers can be observed. PP mesh fibers treated for 24 h (Figure 2e,f) show gamut and thin layer of PDA coating, but have rougher surfaces and white patches along the whole sphere of fibers. In the past, carbon fibers have been successfully coated with PDA [33]. Herein, we also successfully obtained surface changes of PP meshes with a thin layer of PDA coating. Thus, it is proved that bio-inspired PDA can coat PP fibers at room temperature.

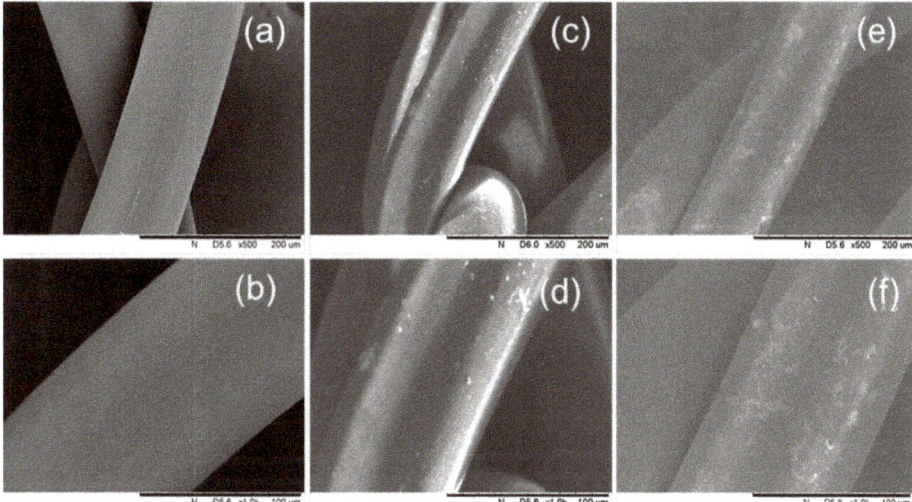

Figure 2. SEM imaged (**a,b**) untreated PP mesh fibers, (**c,d**) PDA treated PP meshes for 12 h, and (**e,f**) PDA treated PP meshes for 24 h.

4.3. Surface Characterization of Polydopamine Functionalized PP Meshes

Figure 3 shows PP control and PDA functionalized PP meshes.

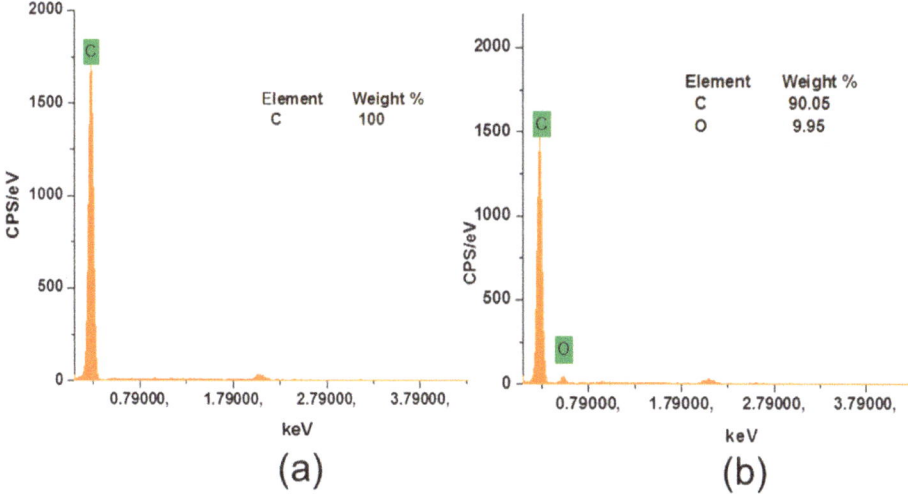

Figure 3. EDX spectra (**a**) weight % of untreated PP mesh fibers and (**b**) after surface functionalization with polydopamine (PDA-24).

It can be observed that untreated PP meshes (Figure 3a) displayed a complete peak height and 100% weight of carbon (C) atom within 0.3 keV. Nevertheless, in Figure 3b, PDA functionalized PP meshes demonstrate an additional peak of the oxygen (O) atom within 0.4 keV, and show a 9.95% oxygen atomic weight increase. However, the carbon (C) atom peak is at a similar position, but the weight % of carbon atom is reduced to 90.05%. Thus, it is proved that PP meshes were successfully functionalized with PDA.

The FTIR (ATR) spectra of polydopamine (PDA) functionalized and untreated PP meshes fibers are shown in Figure 4. The untreated PP meshes demonstrate peaks at 2951 cm^{-1}, 2917 cm^{-1}, 1451 cm^{-1}, and 1377 cm^{-1} [36,37]. Nevertheless, PDA functionalized PP meshes for 12 h exhibit an additional peak of hydroxyl (OH) at 3220 cm^{-1}. Moreover, identical vibration peak bands can be observed at 1624 cm^{-1} (amide I) and 1535 cm^{-1} (amide II). Thus, these peaks may be due to the C=O stretching vibrations and C-N stretching and N-H bending. Furthermore, PDA functionalized PP meshes for 24 h also show similar structural results and vibration peaks (1624 cm^{-1}, 1535 cm^{-1}) were observed at a similar wavenumber cm^{-1}.

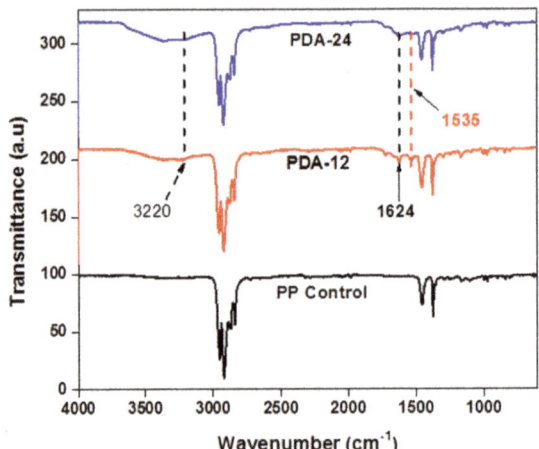

Figure 4. FTIR (ATR) spectra without treatment (PP control), PP meshes functionalized with polydopamine (PDA) for 12 h and 24 h.

4.4. Structural and Thermal Properties

Figure 5a displays the crystal structure of untreated and dopamine functionalized PP meshes. The PP control shows five peak (14.20, 17.13, 18.90, 21.41, and 25) lattices within 2θ [38]. Thus, polydopamine (PDA) functionalized PP fibers maintained same peaks. The crystallinity of PP untreated, PDA treated for 12 h (PDA-12), and PDA treated for 24 h (PDA-24) was 61.2%, 61.31%, and 61.42%, respectively. Therefore, it can be summarized that PDA treatment had no significant effect on the crystallinity of PP fibers.

Figure 5b shows the thermal properties of PP meshes before and after polydopamine treatment. The melting temperature of untreated, and polydopamine functionalized (PDA-12) and (PDA-24) samples were 147.8 °C, 147.6 °C, and 147.9 °C, respectively. It can be noticed that there was no significant change in the melting temperature of treated and untreated mesh samples. Thus, thermal properties before and after treatment were almost similar. The reason for this may be that dopamine functionalized PP meshes had a very thin layer which may not have had a significant impact on the thermal properties.

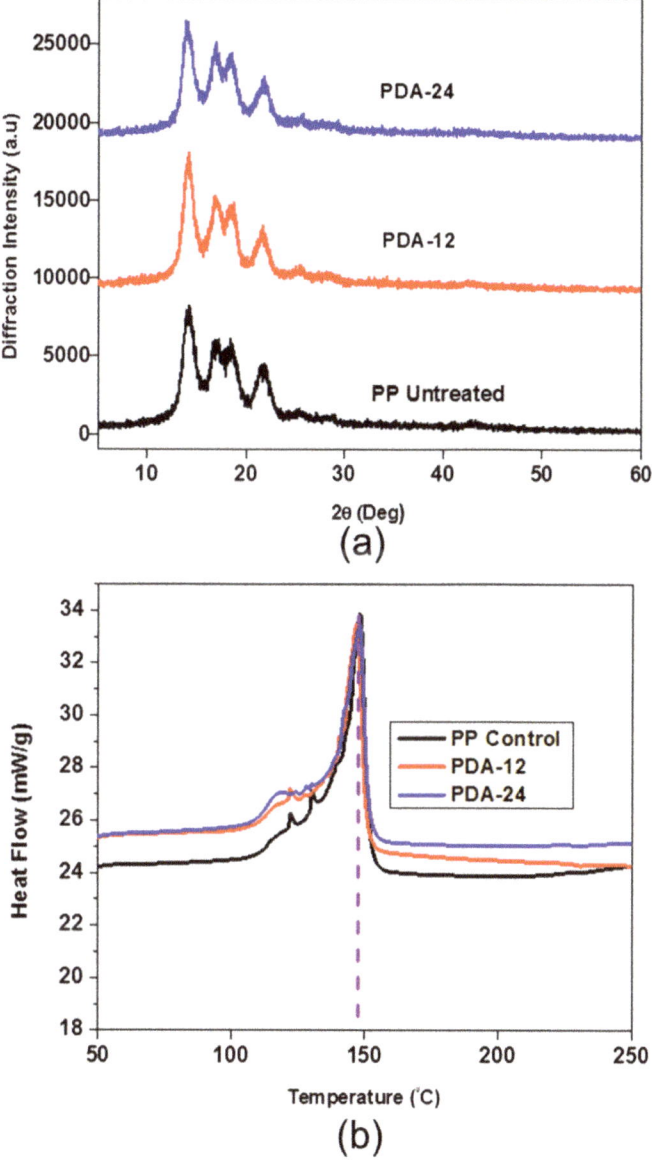

Figure 5. Structural and thermal properties of polydopamine functionalized and untreated PP meshes (**a**) XRD patterns; (**b**) DSC of PP control and PDA modified samples.

4.5. Water Contact Angle

The water contact angles of polydopamine (PDA) functionalized and untreated PP meshes were revealed (Figure 6) by the sessile drop method. As presented in Figure 6A, the water contact angle of PP before surface functionalization was 138.9°, but after 12 h treatment (PDA-12) in dopamine solution, the contact angle decreased to 90.7°.

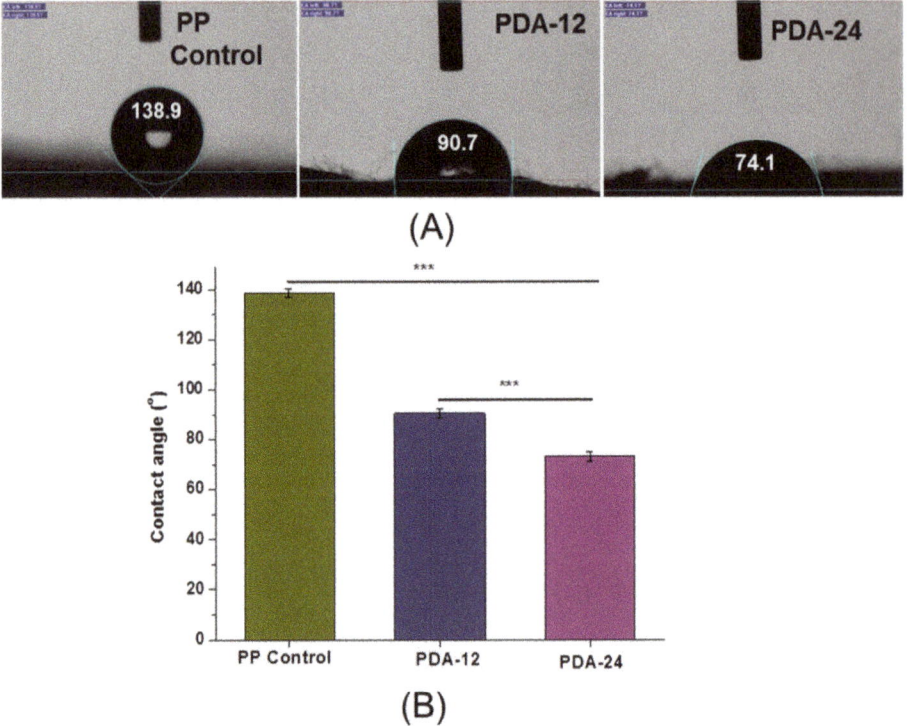

Figure 6. (**A**) Water contact angle drops (a) PP untreated, (b) PDA-12, and (c) PDA-24. (**B**) Average water contact angle of untreated and polydopamine treated PP meshes.

Moreover, the contact angle of PP fabric treated with PDA for 24 h (PDA-24) dramatically decreased up to 74.1°. Thus, an average contact angle difference of untreated (Figure 6B) to PDA-12 and PDA-24 was 34.65% and 47.22%, respectively. However, these results of the contact angle of PDA functionalization are in accord with a recently published paper on polydopamine coating [31]. The reason for contact angle reduction is mainly based on the surface amount of PDA on the PP surfaces. As the functionalization efficiency increased, the contact angle decreased. This is due to the fact that PP was functionalized with PDA and many hydrophilic groups (OH) are present on the surfaces of PP fibers.

5. Conclusions

Polypropylene mesh devices were successfully surface functionalized with polydopamine (PDA). FTIR results evidenced that bio-inspired PDA functionalized the PP mesh surfaces. It was proved by XRD patterns and thermal properties (DSC) that there was no impact of PDA surface functionalization on PP meshes. Thus, the melting temperature and their structure were very similar before and after surface functionalization.

Moreover, surface functionalized PP devices were found to be highly hydrophilic, which is an advantage of PP mesh hernia devices as they absorb soluble antimicrobial drugs during mesh implantation. Thus, PP meshes modified with bio-inspired polydopamine could be a valuable addition to hernia mesh implantation.

Author Contributions: N.S. and F.W. conceived and designed the experiments; N.S. and X.S. performed the experiments; M.P., A.H., and N.S. analyzed the data; N.S. and L.W. wrote the paper.

Funding: This research was funded by 111 project "Biomedical Textile Material Science and Technology" (grant No. B07024), the National Key Research and Development Program of China (Grant No. 2016YFB0303300-03), and the Fundamental Research Funds for the Central Universities (Grant No. 17D110111 & 2232018G-01).

Conflicts of Interest: The authors declare no conflict of interest.

References

1. Baylon, K.; Rodriguez-Camarillo, P.; Elias-Zuniga, A.; Diaz-Elizondo, J.A.; Gilkerson, R.; Lozano, K. Past, Present and Future of Surgical Meshes: A Review. *Membranes* **2017**, *7*, 47. [CrossRef] [PubMed]
2. Biondo-Simoes, M.L.; Carvalho, L.B.; Conceicao, L.T.; Santos, K.B.; Schiel, W.A.; Arantes, M.; Silveira, T.D.; Magri, J.C.; Gomes, F.F. Comparative study of Polypropylene versus Parietex composite(R), Vicryl(R) and Ultrapro(R) meshes, regarding the formation of intraperitoneal adhesions. *Acta Cir. Bras.* **2017**, *32*, 98–107. [CrossRef] [PubMed]
3. Miao, L.; Wang, F.; Wang, L.; Zou, T.; Brochu, G.; Guidoin, R. Physical Characteristics of Medical Textile Prostheses Designed for Hernia Repair: A Comprehensive Analysis of Select Commercial Devices. *Materials* **2015**, *8*, 8148–8168. [CrossRef] [PubMed]
4. Kalaba, S.; Gerhard, E.; Winder, J.S.; Pauli, E.M.; Haluck, R.S.; Yang, J. Design Strategies and Applications of Biomaterials and Devices for Hernia Repair. *Bioact. Mater.* **2016**, *1*, 2–17. [CrossRef] [PubMed]
5. Sanbhal, N.; Miao, L.; Xu, R.; Khatri, A.; Wang, L. Physical structure and mechanical properties of knitted hernia mesh materials: A review. *J. Ind. Text.* **2017**. [CrossRef]
6. Guillaume, O.; Perez-Tanoira, R.; Fortelny, R.; Redl, H.; Moriarty, T.F.; Richards, R.G.; Eglin, D.; Petter Puchner, A. Infections associated with mesh repairs of abdominal wall hernias: Are antimicrobial biomaterials the longed-for solution? *Biomaterials* **2018**, *167*, 15–31. [CrossRef]
7. Kokotovic, D.; Bisgaard, T.; Helgstrand, F. Long-term Recurrence and Complications Associated With Elective Incisional Hernia Repair. *JAMA* **2016**, *316*, 1575–1582. [CrossRef]
8. Kulaga, E.; Ploux, L.; Balan, L.; Schrodj, G.; Roucoules, V. Mechanically Responsive Antibacterial Plasma Polymer Coatings for Textile Biomaterials. *Plasma Process. Polym.* **2014**, *11*, 63–79. [CrossRef]
9. Knetsch, M.L.W.; Koole, L.H. New Strategies in the Development of Antimicrobial Coatings: The Example of Increasing Usage of Silver and Silver Nanoparticles. *Polymers* **2011**, *3*, 340–366. [CrossRef]
10. Deeken, C.R.; Lake, S.P. Mechanical properties of the abdominal wall and biomaterials utilized for hernia repair. *J. Mech. Behav. Biomed. Mater.* **2017**, *74*, 411–427. [CrossRef]
11. Poussier, M.; Deneve, E.; Blanc, P.; Boulay, E.; Bertrand, M.; Nedelcu, M.; Herrero, A.; Fabre, J.M.; Nocca, D. A review of available prosthetic material for abdominal wall repair. *J. Visc. Surg.* **2013**, *150*, 52–59. [CrossRef]
12. Coda, A.; Lamberti, R.; Martorana, S. Classification of prosthetics used in hernia repair based on weight and biomaterial. *Hernia* **2011**, *16*, 9–20. [CrossRef]
13. Shankaran, V.; Weber, D.J.; Reed, R.L., 2nd; Luchette, F.A. A review of available prosthetics for ventral hernia repair. *Ann. Surg.* **2011**, *253*, 16–26. [CrossRef] [PubMed]
14. Greca, F.H.; de Paula, J.B.; Biondo-Simões, M.L.; da Costa, F.D.; da Silva, A.P.; Time, S.; Mansur, A. The influence of differing pore sizes on the biocompatibility of two polypropylene meshes in the repair of abdominal defects. *Hernia* **2001**, *5*, 59–64. [PubMed]
15. Hazebroek, E.J.; Ng, A.; Yong, D.H.; Berry, H.; Leibman, S.; Smith, G.S. Evaluation of lightweight titanium-coated polypropylene mesh (TiMesh) for laparoscopic repair of large hiatal hernias. *Surg. Endosc.* **2008**, *22*, 2428–2432. [CrossRef] [PubMed]
16. Jerabek, J.; Novotny, T.; Vesely, K.; Cagas, J.; Jedlicka, V.; Vlcek, P.; Capov, I. Evaluation of three purely polypropylene meshes of different pore sizes in an onlay position in a New Zealand white rabbit model. *Hernia* **2014**, *18*, 855–864. [CrossRef] [PubMed]
17. Labay, C.; Canal, J.M.; Modic, M.; Cvelbar, U.; Quiles, M.; Armengol, M.; Arbos, M.A. Antibiotic-loaded polypropylene surgical meshes with suitable biological behaviour by plasma functionalization and polymerization. *Biomaterials* **2015**, *71*, 132–144. [CrossRef]
18. Mazaki, T.; Mado, K.; Masuda, H.; Shiono, M. Antibiotic prophylaxis for the prevention of surgical site infection after tension-free hernia repair: A Bayesian and frequentist meta-analysis. *J. Am. Coll. Surg.* **2013**, *217*, 788–801. [CrossRef]

19. Mazaki, T.; Mado, K.; Masuda, H.; Shiono, M.; Tochikura, N.; Kaburagi, M. A randomized trial of antibiotic prophylaxis for the prevention of surgical site infection after open mesh-plug hernia repair. *Am. J. Surg.* **2014**, *207*, 476–484. [CrossRef]
20. Rosen, M.J. Polyester-based mesh for ventral hernia repair: Is it safe? *Am. J. Surg.* **2009**, *197*, 353–359. [CrossRef]
21. Bringman, S.; Conze, J.; Cuccurullo, D.; Deprest, J.; Junge, K.; Klosterhalfen, B.; Parra-Davila, E.; Ramshaw, B.; Schumpelick, V. Hernia repair: The search for ideal meshes. *Hernia* **2010**, *14*, 81–87. [CrossRef] [PubMed]
22. Brown, C.N.; Finch, J.G. Which mesh for hernia repair? *Ann. R. Coll. Surg. Engl.* **2010**, *92*, 272–278. [CrossRef] [PubMed]
23. Falagas, M.E.; Kasiakou, S.K. Mesh-related infections after hernia repair surgery. *Clin. Microbiol. Infect.* **2005**, *11*, 3–8. [CrossRef] [PubMed]
24. Bilsel, Y.; Abci, I. The search for ideal hernia repair; mesh materials and types. *Int. J. Surg.* **2012**, *10*, 317–321. [CrossRef] [PubMed]
25. Harth, K.C.; Rosen, M.J.; Thatiparti, T.R.; Jacobs, M.R.; Halaweish, I.; Bajaksouzian, S.; Furlan, J.; von Recum, H.A. Antibiotic-releasing mesh coating to reduce prosthetic sepsis: An in vivo study. *J. Surg. Res.* **2010**, *163*, 337–343. [CrossRef] [PubMed]
26. Faulk, D.M.; Londono, R.; Wolf, M.T.; Ranallo, C.A.; Carruthers, C.A.; Wildemann, J.D.; Dearth, C.L.; Badylak, S.F. ECM hydrogel coating mitigates the chronic inflammatory response to polypropylene mesh. *Biomaterials* **2014**, *35*, 8585–8595. [CrossRef] [PubMed]
27. Perez-Kohler, B.; Fernandez-Gutierrez, M.; Pascual, G.; Garcia-Moreno, F.; San Roman, J.; Bellon, J.M. In vitro assessment of an antibacterial quaternary ammonium-based polymer loaded with chlorhexidine for the coating of polypropylene prosthetic meshes. *Hernia* **2016**, *20*, 869–878. [CrossRef]
28. Bellon, J.M.; Rodriguez, M.; Garcia-Honduvilla, N.; Pascual, G.; Bujan, J. Partially absorbable meshes for hernia repair offer advantages over nonabsorbable meshes. *Am. J. Surg.* **2007**, *194*, 68–74. [CrossRef]
29. Patel, H.; Ostergard, D.R.; Sternschuss, G. Polypropylene mesh and the host response. *Int. Urogynecol. J.* **2012**, *23*, 669–679. [CrossRef]
30. Perez-Kohler, B.; Bayon, Y.; Bellon, J.M. Mesh Infection and Hernia Repair: A Review. *Surg. Infect.* **2016**, *17*, 124–137. [CrossRef]
31. Liu, Y.; Fang, Y.; Qian, J.; Liu, Z.; Yang, B.; Wang, X. Bio-inspired polydopamine functionalization of carbon fiber for improving the interfacial adhesion of polypropylene composites. *RSC Adv.* **2015**, *5*, 107652–107661. [CrossRef]
32. Dung The Nguyen, T.H.A.N. Surface modification of polyamide thin film composite membrane by coating of titanium dioxide nanoparticles. *J. Sci.* **2016**, *1*, 468–475. [CrossRef]
33. Lei, Z.; Yu, Z.; Xi-Cang, R.; Cheng-Yun, N.; Guo-Xin, T.; Ying, T.A.N. Bioinspired Polydopamine Functionalization of Titanium Surface for SilverNanoparticles Immobilization with Antibacterial Property. *J. Inorg. Mater.* **2014**, *29*, 1320. [CrossRef]
34. Zhang, R.-X.; Leen, B.; Liu, T.-Y.; Luis Alconero, P.; Wang, X.-L.; Van der Bruggen, B. Remarkable Anti-Fouling Performance of TiO$_2$-Modified TFC Membranes with Mussel-Inspired Polydopamine Binding. *Appl. Sci.* **2017**, *2017*, 81. [CrossRef]
35. Wei, Q.; Haag, R. Universal polymer coatings and their representative biomedical applications. *Mater. Horizons* **2015**, *2*, 567–577. [CrossRef]
36. Nava-Ortiz, C.A.; Alvarez-Lorenzo, C.; Bucio, E.; Concheiro, A.; Burillo, G. Cyclodextrin-functionalized polyethylene and polypropylene as biocompatible materials for diclofenac delivery. *Int. J. Pharm.* **2009**, *382*, 183–191. [CrossRef]
37. Sarau, G.; Bochmann, A.; Lewandowska, R.; Christianse, S. From Micro– to Macro–Raman Spectroscopy: Solar Silicon for a Case Study. In *Advanced Aspects of Spectroscopy*; InTech: Vienna, Austria, 2012.
38. Lin, J.-H.; Pan, Y.-J.; Liu, C.-F.; Huang, C.-L.; Hsieh, C.-T.; Chen, C.-K.; Lin, Z.-I.; Lou, C.-W. Preparation and Compatibility Evaluation of Polypropylene/High Density Polyethylene Polyblends. *Materials* **2015**, *8*, 8850–8859. [CrossRef]

© 2019 by the authors. Licensee MDPI, Basel, Switzerland. This article is an open access article distributed under the terms and conditions of the Creative Commons Attribution (CC BY) license (http://creativecommons.org/licenses/by/4.0/).

Communication

Solvent-Free Reactive Vapor Deposition for Functional Fabrics: Separating Oil–Water Mixtures with Fabrics

Nongyi Cheng, Kwang-Won Park and Trisha L. Andrew *

Department of Chemistry, University of Massachusetts Amherst, Amherst, MA 01003, USA; chengnongyi@outlook.com (N.C.); bryan.kwangwon.park@gmail.com (K.-W.P.)
* Correspondence: tandrew@umass.edu; Tel.: +1-413-545-1651

Received: 18 November 2018; Accepted: 22 December 2018; Published: 1 January 2019

Abstract: A facile, solvent-minimized approach to functionalize commercial raw fabrics is described. Reactive vapor deposition of conjugated polymers followed by post-deposition functionalization transforms common, off-the-shelf textiles into distinctly hydrophobic or superhydrophilic materials. The fabric coatings created by reactive vapor deposition are especially resistant to mechanical and solvent washing, as compared to coatings applied by conventional, solution-phase silane chemistries. Janus fabrics with dissimilar wettability on each face are also easily created using a simple, three-step vapor coating process, which cannot be replicated using conventional solution phase functionalization strategies. Hydrophobic fabrics created using reactive vapor deposition and post-deposition functionalization are effective, reusable, large-volume oil–water separators, either under gravity filtration or as immersible absorbants.

Keywords: reactive vapor deposition; hydrophobic; Janus material; absorbant; separation

---

1. Introduction

Industrial water contamination is a pernicious worldwide problem that wreaks tremendous, long-lasting environmental damage and particularly cripples developing economies. Numerous research efforts are dedicated to developing materials for cleaning up oil-contaminated water [1]. Conventional separation materials such as dispersants, solidifiers, and booms and skimmers suffer from high manufacturing costs, secondary pollution, frequent fouling and low separation efficiency [2,3]. Due to these limitations, membranes with special wettability (superhydrophobicity and oleophilicity) have received intense recent attention. Nanostructured hydrophobic materials, such as polymer membranes and metal meshes, have been reported for oil–water separation, but complex and non-scalable fabrication process are typically encountered, in addition to the frequent need for high-cost, fine chemicals during synthesis [4–8].

In theory, natural fabrics such as cotton and wool are excellent candidates for contaminant absorbing/filtering materials because of their naturally-occurring microscale roughness and inherent porosity. Natural fabrics are also outstanding decontaminants because of their low cost and reusability [9]. Methods of functionalizing commonly-available fabrics for water decontamination include surface modification with inorganic nanoparticles [10–12], silane treatment [13], graft polymerization [14,15] and polymer lamination [16].

However, paradoxically, the industrial processes used to manufacture textiles are, themselves, major contributors to global water pollution [17]. Conventional textile and garment production is water-intensive, consuming approximately 700 gallons of fresh water to produce a T-shirt and 1800 gallons of fresh water to produce a pair of jeans. In 2015, the World Bank estimated that 20% of global water pollution was caused by textile processing [17]. Treating such large volumes of waste water is time and energy-intensive, and expensive. Moreover, the inorganic nanoparticles that are widely used to modify the surface of textiles are known to cause unique environmental damage and human health issues upon inevitable leaching into water sources [18,19]. Therefore, alternative, less solvent-intensive approaches to process and dye textiles are sorely needed.

Here, we report a facile, solvent-minimized approach to functionalize commercial raw fabrics. Reactive vapor deposition of conjugated polymers [20–23] followed by post-deposition functionalization transforms common, off-the-shelf textiles into distinctly hydrophobic or superhydrophilic materials that are effective filters and/or absorbants to remove oil-based contaminants from fresh water, even in the presence of surfactants. The vapor coating method reported herein has the potential to significantly curtail the solvent use associated with mainstream textile manufacturing processes while also creating functional fabrics that can decontaminate polluted water.

2. Materials and Methods

All chemicals were purchased from Millipore Sigma and used without further purification. Fabrics were purchased from fabric stores and used without cleaning.

The reactive vapor deposition of the monomer 3,4-(hydroxymethyl)ethylenedioxy-thiophene (HMEDOT) to create poly(3,4-(hydroxymethyl)ethylenedioxy-thiophene) (PHMEDOT) films on various substrates was carried out in a custom-built, tubular vacuum chamber (Figure 1) whose design and operating principles were previously described [20]. The process pressure was maintained close to 150 mTorr during deposition. The crucible containing the oxidant, $FeCl_3$, was placed 5 inches from the monomer vapor inlet. The substrates, glass slides and fabrics, were placed between the monomer and oxidant vapor source in the tube with only one side exposed to the vapor. First, with the monomer valve closed, the oxidant $FeCl_3$, the substrates, and the monomer HMEDOT, were heated at 170 °C, 95 °C, and 120 °C, respectively, for 8 min. Second, the monomer valve was opened, and the polymer films started to form in the middle of the two vapor sources. The polymerization was halted by closing the valve after 20 min. Third, the polymer films were cooled to room temperature and rinsed with solutions in the following sequence: methanol (20 min), 1 M HCl solution (15 min), and methanol (20 min). The samples were dried in air. Chemical characterization of PHMEDOT films is previously reported [22].

PHMEDOT-coated fabrics, glass slides or pristine fabrics were placed in a sealed glass container together with a few small droplets (approximately 0.1–0.5 mL) of neat trichloro-(1H,1H,2H,2H-perfluorooctyl)-silane and the container was placed in an oven held at 60 °C overnight. Then, the samples were rinsed with ethanol for 20 min (2×) and dried in a vacuum oven at 50 °C for 2 h.

Commercial, untreated fabrics were immersed in a 1 wt.% solution of trichloro(1H,1H,2H,2H-perfluorooctyl)silane in Tetrahydrofuran (THF) and the solution was maintained at 50 °C overnight. Then, the samples were extracted from the solution, rinsed with fresh THF for 20 min (2×) and dried in a vacuum oven at 50 °C for 2 h.

3. Results and Discussion

Fabrics were investigated as potential filters/absorbants for three reasons. First, fabrics are already manufactured in a high volume and can be recycled. This is important for cleaning high-volume pollution events, as a large quantity of contaminant absorbing/separating materials are needed. Second, fabrics have highly-textured surfaces with a large surface area, which should afford unmatched contaminant separation from dilute media [24]. Third, fabrics are naturally porous materials with nanoscale and microscale pores that can be tuned by judiciously choosing the constituent fiber (cotton, silk, wool) and weave or knit pattern, affording numerous experimental handles with which to systematically study and control contaminant transport, sequestering and/or capture.

However, commonly-available, untreated fabrics indiscriminately absorb any liquid to which they are exposed and are, therefore, not capable of selectively removing contaminants from aqueous mixtures. For example, when an oil/water bilayer is gravity-filtered through a commercially-available raw cotton fabric, the fabric absorbs a portion of both the oil and water, and minimal overall separation is effected (see video in Supplementary Materials). Here, reactive vapor deposition [20–23] was investigated as a solvent-minimized method to functionalize the surface of untreated textiles and create fabric filters and/or absorbants for water decontamination.

Figure 1 illustrates the stepwise process used to functionalize commercial, raw fabrics. First, off-the-shelf textiles were coated with PHMEDOT in a custom-built reactor that allows rough and/or fragile substrates to be nondestructively coated with a variety of conjugated polymer films [20]. The PHMEDOT coating changed the color of the starting fabric into a dark blue and created a superhydrophilic surface. For all the fabrics reported in this work, the PHMEDOT coating thickness was 300 nm (as measured on a flat silicon test silicon coupon that was coated concomitantly with the fabrics [20–22]). As previously reported, the PHMEDOT coating on all investigated fabrics, irrespective of weave density, was conformal and the intrinsic porosity and hand-feel of the starting fabrics was maintained [21,22]. Next, the fabrics were exposed to vapors of trichloro(1H,1H,2H,2H-perfluorooctyl)silane [25] to make a hydrophobic surface labelled F-PHMEDOT. No color change was observed after this post-deposition functionalization. The total weight of all investigated fabrics increased by less than 0.1% relative to the pristine samples after the two-step coating process.

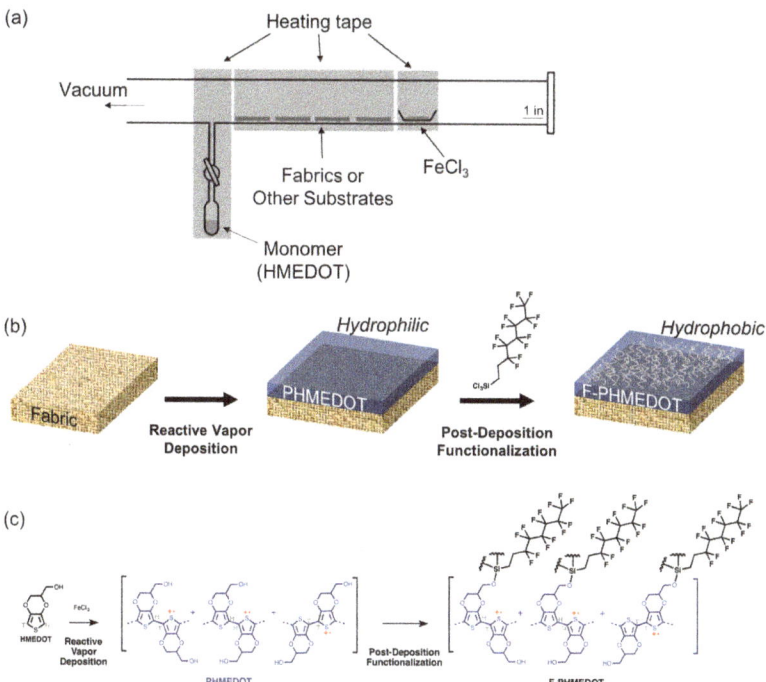

Figure 1. (a) Schematic of custom-built reactor used to coat untreated fabrics via reactive vapor deposition. (b) Illustration of the process to tune the wettability of off-the-shelf fabrics using reactive vapor deposition and post deposition functionalization. (c) Chemical structures of poly(3,4-(hydroxymethyl)ethylenedioxy-thiophene) (PHMEDOT) and F-PHMEDOT polymer coatings on fabrics.

As the example of an untreated cotton fabric, the pristine, as-purchased sample was naturally hydrophilic, with an apparent contact angle of 62.8° (Figure 2a), but this pristine fabric sample slowly absorbed the water droplet over approximately 10 min. After coating with PHMEDOT, the fabric surface became superhydrophilic. An apparent water contact angle was unmeasureable for PHMEDOT-coated fabrics because the water droplet was immediately absorbed by the fabric upon contact (Figure 2b). After post-deposition functionalization to transform PHMEDOT into F-PHMEDOT, the surface of this same fabric became hydrophobic, with an apparent water contact angle of 140.7° (Figure 2c). Therefore, reactive vapor deposition can be used in combination with post-deposition functionalization to create either hydrophilic or hydrophobic fabrics.

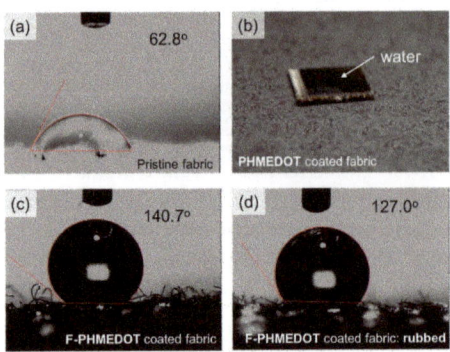

Figure 2. (**a**) Apparent water contact angle of a pristine, tight-woven, raw cotton fabric. (**b**) Photograph of a tight-woven, raw cotton fabric after vapor coating with PHEMDOT, showing the extreme hydrophilicity of the fabric surface. A water droplet placed on this coated fabric surface is immediately absorbed into the fabric. (**c,d**) Apparent water contact angles of (**c**) a tight-woven cotton fabric after vapor coating with F-PHMEDOT, and (**d**) the same vapor-coated fabric after mechanical rubbing.

The hydrophobic surface of the F-PHMEDOT-coated fabric is mechanically and chemically stable (Figure 2d, Figure 3). In our previous report, we proved that vapor-deposited conjugated polymer films on fabrics were remarkably stable to bending/folding, rubbing, dry ironing and cold laundering [21]. Here, we tested the stability of the post-deposition functionalized F-PHMEDOT coating by comparing the apparent water contact angle of F-PHMEDOT-coated fabrics before and after 500 times of mechanical rubbing and solvent (dichloromethane) washing (Figure 3). Further, the ruggedness of our vapor-deposited F-PHMEDOT coating on fabrics was compared to that of a hydrophobic coating created using conventional, solution-phase methods. The control sample was made by soaking the pristine fabric in 1 wt.% solution of trichloro(1H,1H,2H,2H-perfluorooctyl)silane in THF overnight. The F-PHMEDOT coated fabrics displayed apparent contact angle changes of less than 10% after mechanical rubbing and solvent washing. While the apparent water contact angle of the control sample remained unchanged after solvent washing, the apparent contact angle decreased by more than 30% after mechanical rubbing and the water droplet was fully absorbed by the fabric after 30 s. Therefore, the vapor-deposited coatings described herein possess superior ruggedness compared to fabric coatings created by conventional solution-processing.

An unmatched feature of reactive vapor deposition is that it can be spatially controlled to allow the selective coating of one face of a prewoven fabric. Mass transport of the reactive radical species created during reactive vapor deposition can be predictably controlled by the chamber pressure, substrate stage temperature, intrinsic adhesive properties of the reactants, and substrate roughness/porosity [21,22,26,27]. For the reactor and deposition conditions used in this report, the weave density and porosity of the substrate textile was found to be the primary variable controlling the area of surface coating. As seen in Figure 4, use of a prewoven fabric with a loose weave pattern (high porosity) created a uniform PHMEDOT coating across all exposed surfaces of the fabric and no uncoated surfaces were noted. In contrast, when a fabric with a tight weave density (low porosity) was subjected to reactive vapor coating in our chamber, two distinct sides of the fabric were observed, one coated (blue) and one uncoated.

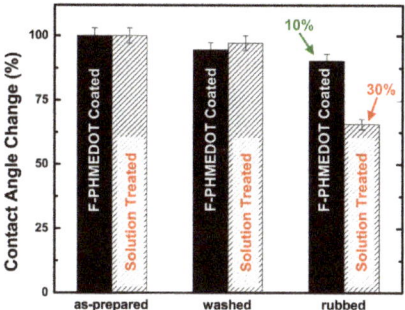

Figure 3. Change in apparent water contact angles of a vapor-coated (F-PHMEDOT) and solution-coated cotton fabric before and after solvent washing and mechanical rubbing.

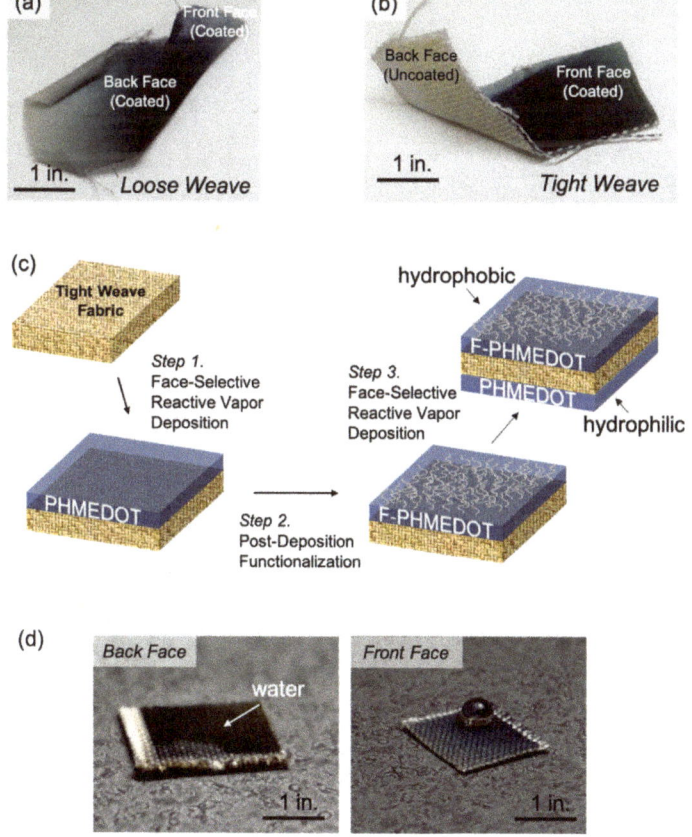

Figure 4. (a) Optical image of a loose weave, untreated cotton fabric vapor coated with PHMEDOT. (b) Optical image of a tight weave, untreated cotton fabric vapor coated with PHMEDOT. (c) Stepwise process for creating Janus fabrics with dissimilar wettability on each face. (d) Optical images of a water droplet on the front and back sides of the Janus fabric.

Taking advantage of this selective coatability feature, tight-woven textiles were transformed into Janus fabrics that displayed two distinct wettabilities on each face/side. By depositing PHMEDOT on the clean, uncoated side of an F-PHMEDOT coated fabric, Janus fabrics with one hydrophilic face and one hydrophobic face were created (Figure 4). Figure 4d shows the distinctive wettability of each face of the Janus fabric. A video demonstrating the dissimilar hydrophilic and hydrophobic properties of each face of the fabric is provided in the Supplementary Materials. Janus fabrics will be useful for separating both oil-in-water emulsions and water-in-oil emulsions [28,29].

Next, the ability of the F-PHMEDOT-coated fabrics to separate oil/water mixtures was evaluated in two ways. The oil phase used here was hexanes-dyed with Oil Red O. In the first method, a two-phase oil/water bilayer was poured into a funnel containing an F-PHMEDOT-coated tight woven cotton fabric as the filter (Figure 5a–c). The two-phase oil water mixture was separated into two distinct components by gravity, without extra applied pressure. The gravity-driven separation process was very fast (2 s) and occurred concomitantly with liquid flow through the fabric filter. A video demonstrating oil/water separation in real time is provided in the Supplementary Materials. The fabric filter remained unfouled even after large volumes of oil/water mixtures (two litres total) were continuously filtered through, indicating that large volume of oil-contaminated water can be purified using this fabric filter.

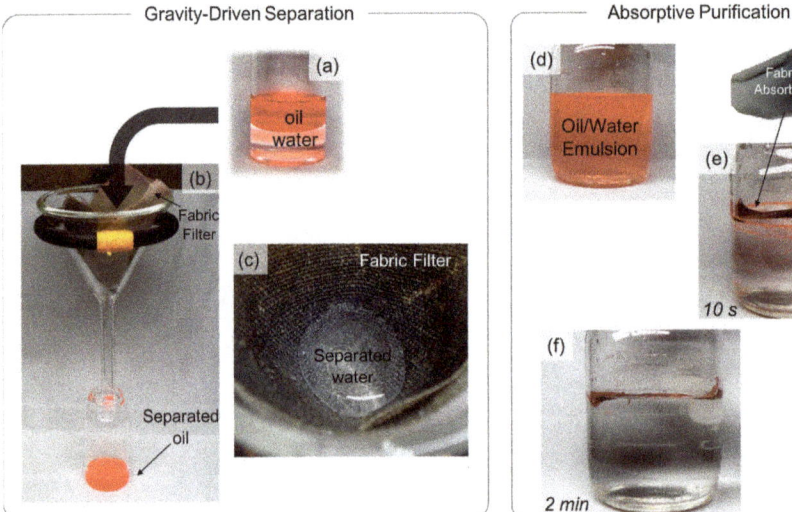

Figure 5. Gravity-driven separation of a two-phase oil/water mixture (**a**) using an F-PHMEDOT coated tight woven cotton fabric as a pass-through filter (**b**). The water phase remains in the fabric filter (**c**) while the oil phase is filtered through and collected (**b**). Removal of oil from a surfactant-stabilized oil/water emulsion (**d**) by immersing an F-PHMEDOT coated loose woven cotton fabric into the emulsion. The fabric absorbant selectively soaks up the oil particles after a short time (**e**) and oil is completely sequestered in the fabric after 2 min (**f**).

In the second method, a surfactant-stabilized oil/water emulsion was clarified (Figure 5d–f). The surfactant-stabilized mixture had an oil to water ratio of 1:100 by volume and was stabilized with commercial hand-washing soap. A 1 inch × 2 inch piece of an F-HMEDOT-coated loose woven cotton fabric was fully immersed in the mixture. Within 10 s, the red-colored oil was absorbed by the fabric and the mixture became clear after 2 min (Figure 5f). The fabric absorbant depicted in Figure 5f could be further re-used to purify another oil/water emulsion after simply wringing it out and air drying; this process was repeated five times without any noticeable failure in the oil-absorbing capacity of the 1 inch × 2 inch piece of fabric.

4. Conclusions

A combination of reactive vapor deposition and post-deposition functionalization affords fabrics with distinctive hydrophobic or superhydrophilic surfaces. The surface coatings created by reactive vapor deposition are especially resistant to mechanical washing, as compared to surface coatings applied by conventional, solution-phase silane chemistries. Janus fabrics with dissimilar wettability on each face can also be created using a simple, three-step vapor coating process, which cannot be replicated by solution-phase methods. The hydrophobic fabric reported herein acts as an effective, reusable, high-volume oil-water separation material, either under gravity filtration (for phase-separated mixtures) or as an immersible absorbant (for surfactant-stabilized oil/water emulsions).

Conventional textile and garment production is water-intensive, consuming approximately 700 gallons of fresh water to produce a T-shirt and 1800 gallons of fresh water to produce a pair of jeans. In 2015, the World Bank estimated that 20% of global water pollution was caused by textile processing [17]. Treating such large volumes of waste water is time and energy-intensive, and expensive. Therefore, alternative, less solvent-intensive approaches to process and dye textiles are needed.

Reactive vapor deposition allows for the solvent-free functionalization of raw, off-the-shelf fabrics and has the potential to significantly curtail the solvent use associated with mainstream textile manufacturing processes [30]. Currently, however, vapor coating methods are not broadly adopted by textile scientists and manufacturers because of the perceived difficulty and high cost of scaling up vapor coating chambers to satisfy the high volume demand of the textile industry. Burgeoning reactor designs for stain-guarding carpets [31] prove that vapor coating methods are indeed conducive to large-scale, high-throughput manufacturability. Nevertheless, further research in the academic sector is needed to optimize reactor designs and coating stages to continuously process and coat large spools of thread. Innovations in vapor phase polymerization chemistries are also required to decrease the duration of each coating cycle (currently 20 min) and eliminate the use of heavy metal oxidants during the coating process.

Supplementary Materials: Videos of oil-water separation in real time. The following are available online at http://www.mdpi.com/2079-6439/7/1/2/s1, Janus Fabric (.mp4), Functionalized Fabric Gravity Filter (.mp4), Ineffective Raw Fabric Filter (.mp4).

Author Contributions: Conceptualization, N.C. and T.L.A.; methodology, N.C., K.-W.P. and T.L.A.; formal analysis, N.C. and K.-W.P.; investigation, N.C. and K.-W.P.; writing—original draft preparation, N.C.; writing—review and editing, T.L.A.; funding acquisition, T.L.A.

Funding: This material is based upon work supported by the National Science Foundation under CHEM MSN 1807743. T.L.A. also gratefully acknowledges support from the David and Lucille Packard Foundation.

Conflicts of Interest: The authors declare no conflict of interest.

References

1. Kintisch, E. An Audacious Decision in Crisis Gets Cautious Praise. *Science* **2010**, *329*, 735–736. [CrossRef] [PubMed]
2. Sarbatly, R.; Krishnaiah, D.; Kamin, Z. A review of polymer nanofibres by electrospinning and their application in oil–water separation for cleaning up marine oil spills. *Mar. Pollut. Bull.* **2016**, *106*, 8–16. [CrossRef] [PubMed]
3. Barry, C. Slick death: Oil-spill treatment kills coral. *Sci. News* **2007**, *172*, 67. [CrossRef]
4. Tang, X.; Si, Y.; Ge, J.; Ding, B.; Liu, L.; Zheng, G.; Luo, W.; Yu, J. In situ polymerized superhydrophobic and superoleophilic nanofibrous membranes for gravity driven oil-water separation. *Nanoscale* **2013**, *5*, 11657–11664. [CrossRef] [PubMed]
5. Feng, L.; Zhang, Z.; Mai, Z.; Ma, Y.; Liu, B.; Jiang, L.; Zhu, D. A Super-Hydrophobic and Super-Oleophilic Coating Mesh Film for the Separation of Oil and Water. *Angew. Chem. Int. Ed.* **2004**, *43*, 2012–2014. [CrossRef] [PubMed]

6. Zhang, F.; Zhang, W.B.; Shi, Z.; Wang, D.; Jin, J.; Jiang, L. Nanowire-Haired Inorganic Membranes with Superhydrophilicity and Underwater Ultralow Adhesive Superoleophobicity for High-Efficiency Oil/Water Separation. *Adv. Mater.* **2013**, *25*, 4192–4198. [CrossRef] [PubMed]
7. Song, J.; Huang, S.; Lu, Y.; Bu, X.; Mates, J.E.; Ghosh, A.; Ganguly, R.; Carmalt, C.J.; Parkin, I.P.; Xu, W.; et al. Self-Driven One-Step Oil Removal from Oil Spill on Water via Selective-Wettability Steel Mesh. *ACS Appl. Mater. Interfaces* **2014**, *6*, 19858–19865. [CrossRef] [PubMed]
8. Wu, J.; Wang, N.; Wang, L.; Dong, H.; Zhao, Y.; Jiang, L. Electrospun Porous Structure Fibrous Film with High Oil Adsorption Capacity. *ACS Appl. Mater. Interfaces* **2012**, *4*, 3207–3212. [CrossRef]
9. Zhang, J.; Seeger, S. Polyester Materials with Superwetting Silicone Nanofilaments for Oil/Water Separation and Selective Oil Absorption. *Adv. Funct. Mater.* **2011**, *21*, 4699–4704. [CrossRef]
10. Zhang, J.; Li, B.; Wu, L.; Wang, A. Facile preparation of durable and robust superhydrophobic textiles by dip coating in nanocomposite solution of organosilanes. *Chem. Commun.* **2013**, *49*, 11509–11511. [CrossRef]
11. Zhang, X.; Geng, T.; Guo, Y.; Zhang, Z.; Zhang, P. Facile fabrication of stable superhydrophobic SiO_2/polystyrene coating and separation of liquids with different surface tension. *Chem. Eng. J.* **2013**, *231*, 414–419. [CrossRef]
12. Zhang, M.; Wang, C.; Wang, S.; Li, J. Fabrication of superhydrophobic cotton textiles for water–oil separation based on drop-coating route. *Carb. Polym.* **2013**, *97*, 59–64. [CrossRef] [PubMed]
13. Zhang, X.; Shi, F.; Niu, J.; Jiang, Y.; Wang, Z. Superhydrophobic surfaces: From structural control to functional application. *J. Mater. Chem.* **2008**, *18*, 621–633. [CrossRef]
14. Qu, M.; Hou, L.; He, J.; Feng, J.; Liu, S.; Yao, Y. Facile process for the fabrication of durable superhydrophobic fabric with oil/water separation property. *Fiber Polym.* **2016**, *17*, 2062–2068. [CrossRef]
15. Deng, B.; Cai, R.; Yu, Y.; Jiang, H.; Wang, C.; Li, J.; Li, L.; Yu, M.; Li, J.; Xie, L.; et al. Laundering Durability of Superhydrophobic Cotton Fabric. *Adv. Mater.* **2010**, *22*, 5473–5477. [CrossRef] [PubMed]
16. Yoo, Y.; You, J.B.; Choi, W.; Im, S.G. A stacked polymer film for robust superhydrophobic fabrics. *Polym. Chem.* **2013**, *4*, 1664–1671. [CrossRef]
17. Scott, A. Cutting out textile pollution. *Chemical & Engineering News*, 19 October 2015; 18–19.
18. Bystrzejewska-Piotrowska, G.; Golimowski, J.; Urban, P.L. Nanoparticles: Their potential toxicity, waste and environmental management. *Waste Manag.* **2009**, *29*, 2587–2595. [CrossRef]
19. Colvin, V.L. The potential environmental impact of engineered nanomaterials. *Nat. Biotechnol.* **2003**, *21*, 1166. [CrossRef]
20. Cheng, N.; Andrew, T.L. Reactive Vapor Deposition of Conjugated Polymer Films on Arbitrary Substrates. *J. Vis. Exp.* **2018**, *131*, e56775. [CrossRef] [PubMed]
21. Zhang, L.; Fairbanks, M.; Andrew, T.L. Rugged Textile Electrodes for Wearable Devices Obtained by Vapor Coating Off-the-Shelf, Plain-Woven Fabrics. *Adv. Funct. Mater.* **2017**, *27*, 1700415. [CrossRef]
22. Cheng, N.; Zhang, L.; Joon Kim, J.; Andrew, T.L. Vapor phase organic chemistry to deposit conjugated polymer films on arbitrary substrates. *J. Mater. Chem. C* **2017**, *5*, 5787–5796. [CrossRef]
23. Bhattacharyya, D.; Howden, R.M.; Borrelli, D.C.; Gleason, K.K. Vapor phase oxidative synthesis of conjugated polymers and applications. *J. Polym. Sci. B Polym. Phys.* **2012**, *50*, 1329–1351. [CrossRef]
24. Adebajo, M.O.; Frost, R.L.; Kloprogge, J.T.; Carmody, O.; Kokot, S. Porous Materials for Oil Spill Cleanup: A Review of Synthesis and Absorbing Properties. *J. Porous Mater.* **2003**, *10*, 159–170. [CrossRef]
25. Genzer, J.; Efimenko, K. Creating Long-Lived Superhydrophobic Polymer Surfaces Through Mechanically Assembled Monolayers. *Science* **2000**, *290*, 2130–2133. [CrossRef] [PubMed]
26. Alf, M.E.; Asatekin, A.; Barr, M.C.; Baxamusa, S.H.; Chelawat, H.; Ozaydin-Ince, G.; Petruczok, C.D.; Sreenivasan, R.; Tenhaeff, W.E.; Trujillo, N.J.; et al. Chemical Vapor Deposition of Conformal, Functional, and Responsive Polymer Films. *Adv. Mater.* **2010**, *22*, 1993–2027. [CrossRef] [PubMed]
27. Zhang, L.; Baima, M.; Andrew, T.L. Transforming Commercial Textiles and Threads into Sewable and Weavable Electric Heaters. *ACS Appl. Mater. Interfaces* **2017**, *9*, 32299–32307. [CrossRef] [PubMed]
28. Gu, J.; Xiao, P.; Chen, J.; Zhang, J.; Huang, Y.; Chen, T. Janus Polymer/Carbon Nanotube Hybrid Membranes for Oil/Water Separation. *ACS Appl. Mater. Interfaces* **2014**, *6*, 16204–16209. [CrossRef]
29. Yun, J.; Khan, F.A.; Baik, S. Janus Graphene Oxide Sponges for High-Purity Fast Separation of Both Water-in-Oil and Oil-in-Water Emulsions. *ACS Appl. Mater. Interfaces* **2017**, *9*, 16694–16703. [CrossRef]

30. Allison, L.; Hoxie, S.; Andrew, T.L. Towards Seamlessly-Integrated Textile Electronics: Methods to Coat Fabrics and Fibers with Conducting Polymers for Electronic Applications. *Chem. Commun.* **2017**, *53*, 7182–7193. [CrossRef]
31. Kovacik, P.; del Hierro, G.; Livernois, W.; Gleason, K.K. Scale-up of oCVD: large-area conductive polymer thin films for next-generation electronics. *Mater. Horiz.* **2015**, *2*, 221–227. [CrossRef]

© 2019 by the authors. Licensee MDPI, Basel, Switzerland. This article is an open access article distributed under the terms and conditions of the Creative Commons Attribution (CC BY) license (http://creativecommons.org/licenses/by/4.0/).

Article

Preparation of Chitosan-Coated Poly(L-Lactic Acid) Fibers for Suture Threads

Daiki Komoto, Ryoka Ikeda, Tetsuya Furuike and Hiroshi Tamura *

Faculty of Chemistry, Materials and Bioengineering, Kansai University, Suita, Osaka 564-8680, Japan; k246956@kansai-u.ac.jp (D.K.); shineeworld99@yahoo.co.jp (R.I.); furuike@kansai-u.ac.jp (T.F.)
* Correspondence: tamura@kansai-u.ac.jp; Tel.: +81-6-6368-1121

Received: 26 September 2018; Accepted: 24 October 2018; Published: 25 October 2018

Abstract: Poly(L-lactic acid) (PLA) is a biodegradable fiber, and a promising material for use in biomedical applications. However, its hydrophobicity, low hydrolyzability, and poor cell adhesion can be problematic in some cases; consequently, the development of improved PLA-based materials is required. In this study, chitosan-coated (CS-coated) PLA was prepared by plasma treatment and the layer-by-layer (LBL) method. Plasma treatment prior to CS coating effectively hydrophilized and activated the PLA surface. The LBL method was used to increase the number of CS and sodium alginate (SA) coating layers by electrostatically superposing alternating anionic and cationic polymers. The prepared fibers were characterized by tensile testing, scanning electron microscopy (SEM), X-ray photoelectron spectroscopy (XPS), nitrogen analysis and degradation testing, which revealed that the 100 W plasma treatment for 60 s was optimum, and that plasma treatment and the LBL method effectively coated CS onto the PLA fibers. The existence or not of a coating on the PLA fiber did not appear to influence the degradation of the fiber, which is ascribable to the extremely thin coating, as evidenced by nitrogen analysis and SEM. The CS-coated PLA fibers were prepared without damaging the PLA surface and can be used in biomaterial applications such as suture threads.

Keywords: PLA fiber; chitosan; sodium alginate; layer-by-layer method; plasma treatment

1. Introduction

Suture threads require properties such as biocompatibility and strength for surgical applications, because the removal of stitches is cumbersome for the doctor, a burden to doctor and patient, and poses a risk of infection. Poly(L-lactic acid) (PLA) fiber is one of the biodegradable suture threads currently in use. PLA is an aliphatic lactic-acid-based polyester, and is well known to be a carbon-neutral material obtained from some plants [1]. PLA has good mechanical properties and characteristics, such as biodegradability, biocompatibility, and lack of toxicity [2–4]. Therefore, PLA is an interesting material for use in applications that include tissue engineering, drug delivery systems, and implants [5–8]. However, the hydrophobicity, low hydrolyzability, and poor cell adhesion exhibited by the PLA surface are problematic in some biomaterial applications [9]. Several techniques have recently been reported to modify PLA surfaces. For example, PLA surfaces can be coated by hydrophilic polymers, such as proteins, chitosan (CS), and sodium alginate (SA) by alkaline, enzymatic, or plasma treatments [10–14].

CS is derived from chitin, which is well known to be the most abundant natural biopolymer [15–17]. In addition, CS is used prominently as a biomaterial because of its numerous excellent properties, which include biocompatibility, biodegradability, and antibacterial and wound-healing abilities [18–21]. CS is a cationic polymer bearing amino (–NH$_2$) groups, a consequence of its glucosamine units, and is soluble in acidic aqueous solutions [22]. SA is also a natural biopolymer composed of (1,4) β-D-mannuronate and α-L-guluronate [23]. SA is extracted from seaweed and used in the food industry as a thickener, stabilizer and gelatinizer, a consequence of its superior properties that include biocompatibility, moisture-retention, high viscosity, and easy gelation [24–26]. SA is the sodium salt of an anionic polymer bearing carboxylate (–COO$^-$) groups, and is soluble in neutral and basic aqueous solution [27].

In this study, we modified the PLA surface with CS and SA coatings to improve its surface properties while maintaining its biodegradability. Plasma treatment and layer-by-layer (LBL) deposition were employed for preparation. To the best of our knowledge, this is the first report of a CS- and SA-coated PLA fiber prepared. Plasma treatment is effective for coating CS onto the PLA surface without damaging the PLA because it hydrophilizes and activates the surface layer. Moreover, the LBL method effectively increases the amount of CS coating. The LBL method has received attention as a coating technique capable of forming multiple layers of ionic polymers on the surface of a material, and many reports have noted that multilayers can be easily prepared via this method on several materials [28]. The prepared multilayers are uniform, thin, hard, adhesive, and slightly soluble; moreover, the surfaces can be readily modified through the formation of polyelectrolyte layers [29,30]. In this study, SA and CS, as anionic and cationic layers, were coated via the electrostatic superposition of alternating polymers onto PLA fibers with these methods. The mechanical properties, morphologies, CS–PLA and SA–CS interactions, nitrogen content, and degradability of the prepared fibers were characterized by tensile testing, scanning electron microscopy (SEM), X-ray photoelectron spectroscopy (XPS), nitrogen analysis, and degradation testing, respectively. The CS-coating-modified PLA will be applicable as suture threads, because the antibacterial and wound-healing abilities of CS reduce the risk of postoperative infections.

2. Materials and Methods

2.1. Materials

CS (FM-80, Mw = 24.8 × 10^3, degree of deacetylation (DDA) = 84.7%) was provided by Koyo Chemical Co., Ltd. (Sakaiminato, Japan). The weight-average molecular weight (Mw) and the DDA of CS were determined by gel permeation chromatography (GPC) and ^{1}H-NMR spectroscopy, respectively. SA (Grad IL-2, viscosity = 44 mPa·s) was provided by Kimica Corporation (Tokyo, Japan). Other chemicals were purchased from FUJIFILM Wako Pure Chemical Corporation (Osaka, Japan), and were used without purification. PLA fiber (167T48) was composed of 48 filaments of diameter 20 μm.

2.2. Methods

The PLA fiber was coiled on a frame and washed twice with 1% Tween 20 solution. The washed PLA was rinsed three times with distilled water and dried at room temperature. A 1% CS solution was prepared as follows. A 2 g sample of CS was dissolved in 1% aqueous acetic acid and freeze-dried to remove excess acetic acid and water. The obtained CS-acetate salt was then dissolved in 200 mL of water. The PLA fiber was protected against acetic-acid damage through the preparation of the 1% CS solution used in this method. A 1% SA solution was prepared by the overnight stirring of 2 g of SA in 200 mL of water. The coating method involving plasma treatment and the LBL method is described below.

To coat them with CS, the PLA-fiber surfaces were activated by plasma treatment (CUTE–MR/R, FEMTO SCIENCE, Hwaseong, Korea) prior to immersion in the above-mentioned CS solution. Plasma treatment was performed under the following conditions: time, 30–1800 s; power, 100–200 W; and oxygen-gas flow, 70 mL/min (Table 1).

The PLA fibers, plasma-treated under a variety of conditions, were immediately immersed in 1% CS solution and stirred for 1 min at room temperature. The PLA fibers were then rinsed three times with distilled water and dried at room temperature. PLA fibers coated with a single layer of CS were obtained in this manner.

To coat with CS and SA using the LBL method, the CS-coated PLA fibers were first immersed in 1% SA solution and stirred for 1 min at room temperature, after which the PLA fibers were rinsed three times with distilled water. CS-coated PLA fibers coated with SA through electrostatic interactions with CS were produced in this manner.

The PLA fibers formed in this manner were then immersed in 1% CS solution and stirred for 1 min at room temperature, after which the PLA fibers were rinsed three times with distilled water to give CS/SA/CS-coated PLA fibers.

Multilayer-coated PLA fibers were produced by repeating these steps. The sequence was repeated in a manner that ensured that the final coating layer was composed of CS. The coating procedure is schematically displayed in Figure 1, while the numbers of the coating layers on the various samples are summarized in Table 2.

Preformed PLA samples were treated under a variety of conditions to investigate the effect of plasma treatment and the LBL method on the surface of the samples. Tables 1 and 2 show the conditions used for the plasma treatment and the LBL method, respectively.

Table 1. Conditions used for the plasma treatment condition of PLA fibers.

Sample	Power (W)	Time (s)
P1	100	60
P2	100	300
P3	100	1800
P4	200	60
P5	200	300
P6	200	1800

Table 2. Conditions (C1–C5) used during the CS coating of the P1 PLA-fiber sample.

Sample	Number of CS Layers	CS Content (%)	Thickness of the CS Layer (nm)
C1	1	0.331	31.5
C2	3	0.377	35.9
C3	5	0.585	55.8
C4	10	0.730	69.7
C5	15	1.083	103.7

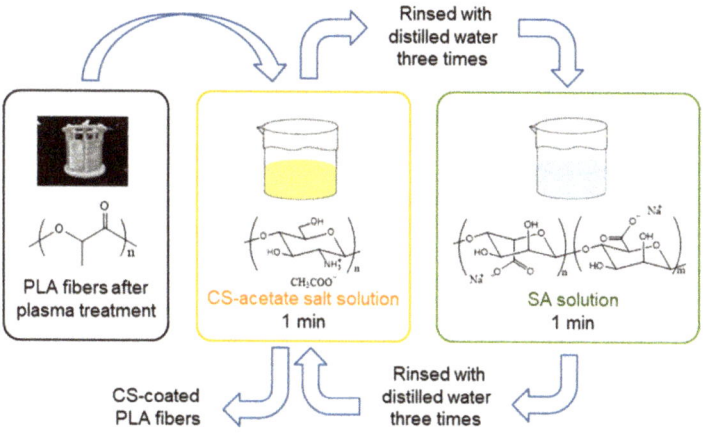

Figure 1. The LBL method used to CS-coat the PLA fibers.

2.3. Tensile Testing

The tensile strengths and strains of the prepared fibers were measured according to the Japanese Industrial Standard (JIS) 1013-8.5 methods using an STA-1150 universal testing machine (Orientec Co., Ltd., Tokyo, Japan). Samples for testing were cut to the appropriate length and attached to paper. The initial sample length was 10 mm, and the stretching rate was 10.0 mm/min. The force at the breaking point was measured as tensile stress, which was converted into tensile strength. Data obtained were the averages of fifteen tests.

2.4. Surface Characterization

The surface morphologies of the samples were investigated by SEM (JSM–6700 microscope, JEOL, Tokyo, Japan). All samples were vacuum-dried overnight and deposited on platinum foil prior to SEM. The CS coatings on the surfaces of the PLA fibers were examined by XPS (ESCA–3400, Kratos Analytical Ltd., Manchester, UK) fitted with a monochromatic Mg–Kα X-ray source, at 10 kV and 20 mA. Spectra in the C 1s (300–276.9 eV), O 1s (524.9–543 eV), and N 1s (410–389.9 eV) binding-energy ranges were acquired. The C 1s peak was separated to three peaks according to the chemical components as follows: hydrocarbon main chain (C–C) at 285.0 eV, ether (C–O) at 286.5 eV, and ester (COO) at 289.2 eV. To compare the amounts of CS in the various samples, the ester (COO/C) and amine (N/C) ratios were calculated using Equations (1) and (2):

$$\text{ester (COO/C)} = \frac{\text{COO peak intensity at 289.2 eV}}{\text{C} - \text{C peak intensity at 285.0 eV}}, \quad (1)$$

$$\text{amino (N/C)} = \frac{\text{C} - \text{NH peak intensity at 400.0 eV}}{\text{C} - \text{C peak intensity at 285.0 eV}}. \quad (2)$$

2.5. Determining the Amount of CS

The Kjeldahl method for the quantitative analysis of nitrogen was used to determine the CS content on the PLA fiber. The samples were vacuum-dried at room temperature overnight prior to analysis.

The sample was degraded in the first step, as follows. A 0.2 g sample was decomposed in a Kjeldahl flask with 10 mL of concentrated H_2SO_4 and 3 g of the catalyst (9:1 K_2SO_4:$CuSO_4$) at 500 °C for 3 h; if colored at this stage, hydrogen peroxide (35 vol%) was added to the solution to decolorize it. The colorless solution obtained in this manner was then heated again at 500 °C for 1 h and then cooled to room temperature, after which 40 mL of deionized water was added.

The degraded sample was steam-distilled and back-titrated in the second step, as follows. The Kjeldahl flask containing the degraded sample was attached to a steam distillation system and 30% w/v aqueous NaOH was added until the solution turned black, which is an indicator of alkalinity. The sample was steam-distilled for 20 min. The free-NH_3-trapping solution was prepared by mixing 4 mL of 0.01 M H_2SO_4, 26 mL of deionized water, and a small amount of ethanolic bromocresol-green/methyl-red indicator solution. Back-titration was performed against a standardized 0.01 M NaOH solution. The CS percentage was calculated using Equations (3) and (4):

$$MW_{CTS} = MW_{GlcN} \times \frac{DDA}{100} + MW_{GlcNAc} \times \frac{100 - DDA}{100}, \quad (3)$$

$$\text{chitosan content (\%)} = \frac{MW_{CTS} \times (C_{H_2SO_4} \times V_{H_2SO_4} - 2 \times C_{NaOH} \times V_{NaOH})}{W}, \quad (4)$$

where MW_{CTS} is the molecular weight of the CS unit (g/mol), MW_{GlcN} is the molecular weight of the glucosamine unit, MW_{GlcNAc} is the molecular weight of the N-acetyl glucosamine unit, C is the concentration of the H_2SO_4 or NaOH solution (mol/L), V is the volume of the H_2SO_4 or NaOH solution (L), and W is the weight of the tested sample (g).

2.6. Degradation Testing in PBS Solution

The PLA fibers were immersed for five months in phosphate-buffered saline (PBS, pH 7.2) solution at 37 °C. At the end of this period, the samples were rinsed three times with distilled water and dried at 35 °C. The degradation rate was calculated from the weight using Equation (5):

$$\text{weight change rate (\%)} = \frac{W_0 - W_1}{W_0} \times 100, \qquad (5)$$

where W_0 and W_1 are the weights of the initial and degraded PLA fibers, respectively.

The degraded PLA fibers were tensile-tested and examined by SEM as described in Sections 2.3 and 2.4.

3. Results and Discussion

3.1. Tensile Testing

3.1.1. Plasma-Treated PLA Fibers

Figure 2 displays the tensile stresses and strains of the various plasma-treated PLA fibers. The average tensile stresses of samples P1–P6 were 359, 345, 334, 378, 359 and 15.7 MPa, respectively. Other than P6, the tensile stresses of the samples were slightly higher than that (312 MPa) of the original PLA fiber; however, the differences were not significant. Fibers treated for longer times exhibited decreased tensile strengths. In particular, the tensile strength of P6 was extremely low, with a stress of about 15.7 MPa, which is ascribable to the observation that P6 melted when exposed to the heat produced during high-power plasma treatment for long times; clearly, PLA is damaged by long plasma treatments. The average strain of each sample was almost identical to that of the original (untreated) fiber. However, P3–P6 showed lower values than P1 and P2, because PLA fibers got damaged under treatment with a high power of 200 W or a long time of 30 min. These results reveal that the strengths and strains of the PLA fibers are not affected by plasma treatment under the appropriate conditions, namely, treatment at low power for short times.

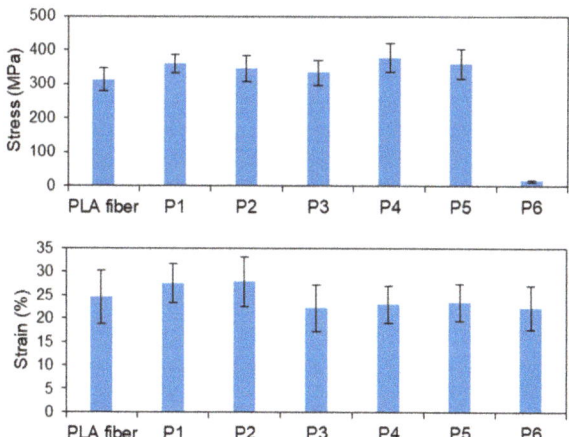

Figure 2. Tensile stresses and strains of the plasma-treated PLA fibers.

3.1.2. CS-Coated PLA Fibers Prepared by the LBL Method

Figure 3 displays the tensile strengths and strains of the CS-coated PLA fibers based on P1, which were chosen based on the results presented in Sections 3.1.1 and 3.2.1 (see below). The average tensile strengths of samples C1–C5 were 346, 341, 328, 325, and 341 MPa, respectively, with C1 exhibiting the same tensile strength and strain as the original PLA fiber. The tensile strengths and strains of C2–C5 were almost identical to that of the original fiber, despite the different number of coating layers. Therefore, all samples prepared by the LBL method have equivalent flexibility comparable with that of the original PLA fiber, which confirms that the LBL method neither damages the PLA fibers nor does it adversely affect their mechanical properties.

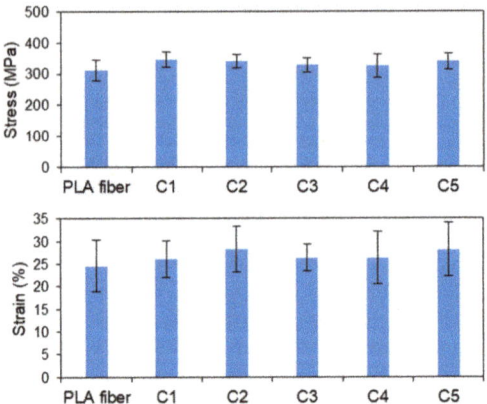

Figure 3. Tensile stresses and strains of the CS-coated PLA fibers based on P1.

3.2. Surface Characterization

3.2.1. Plasma-Treated PLA Fibers

SEM images of the plasma-treated PLA fibers (P1–P6) are displayed in Figure 4. P1 exhibited a smooth surface similar to that of the original PLA fiber. The surfaces of the PLA fibers plasma-treated for longer times were increasingly rough compared with that of P1, especially P6. The fiber diameters of the P1–P5 samples were about 19 µm, while that of P6 was about 74 µm, a result of its partial melting during plasma treatment, as previously discussed; this change in diameter resulted in a decrease in tensile strength. We confirmed that the surfaces of the PLA fibers were damaged by plasma treatment at high power for long times. Wan et al. [31] reported a similar behavior, in that the modifying depth on PLA increased with increasing plasma treatment time. In addition, Ding et al. [12] reported that the amount of coated-CS on PLA by plasma treatment did not change even for treatment durations exceeding 1 min. Consequently, the plasma treatment conditions used to prepare the P1 fibers were determined to be optimum for their further coating with CS; thus, all CS-coated PLA fibers were prepared from P1.

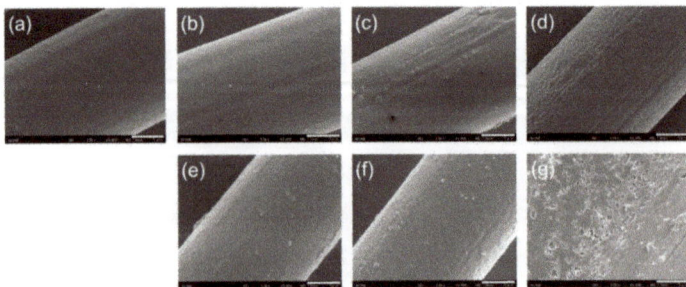

Figure 4. SEM images of plasma-treated PLA fibers: (a) untreated PLA fiber; (b) P1; (c) P2; (d) P3; (e) P4; (f) P5; and (g) P6 (magnification: 5000×, scale bar: 5 µm).

3.2.2. CS-Coated PLA Fibers Prepared by the LBL Method

SEM images of the CS-coated PLA fibers are displayed in Figure 5. C1 exhibited a smooth surface similar to that of the original (untreated) PLA fiber. The surfaces of the remaining fibers became increasingly rough with increasing numbers of coating layers, especially C4 and C5. However, the CS-coated fibers were confirmed to be uniformly coated with CS, as determined by the attachment of each fiber, the presence of CS, and partial CS coatings on the PLA surfaces. In addition, the diameters of the fibers were unchanged following CS coating, despite the PLA surfaces being coated with 30 layers of CS and SA in the case of C5, which confirms that the CS and SA layers are very thin and uniform.

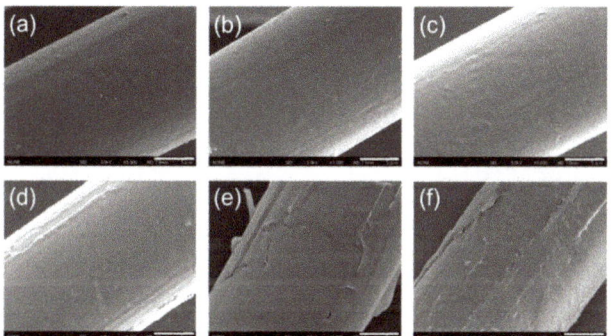

Figure 5. SEM images of CS-coated PLA fibers: (**a**) untreated PLA fiber; (**b**) C1; (**c**) C2; (**d**) C3; (**e**) C4; and (**f**) C5 (magnification: 5000×, scale bar: 5 μm).

The nitrogen-to-carbon (N/C) ratios of the CS-coated PLA fibers were determined by XPS, the results of which are displayed in Figure 6. The amount of CS coating on the PLA surface was determined by measuring the nitrogen and carbon contents, as amide groups are present only in the structure of CS and not in the PLA. The nitrogen content of the original PLA fiber was also determined because a plasticizer is usually added during the preparation of PLA. However, the plasticizer content does not influence other data because the PLA surface was determined to have an N/C ratio of only 0.005 and is covered by the CS layer. The C1 sample exhibited a higher N/C ratio compared to that of the original PLA fiber. While the N/C ratios were higher in C2–C5 than C1, these samples all exhibited similar N/C ratios. This observation reveals that the LBL method produces homogeneous CS coatings, since XPS examines the top surface and the C2 sample already contains a homogeneous CS layer. These results suggest that plasma treatment and the LBL method effectively coat the PLA fibers with CS.

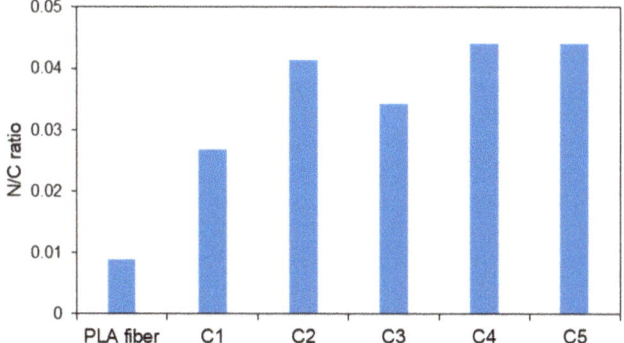

Figure 6. Nitrogen-to-carbon (N/C) ratios of the CS-coated PLA fibers determined by XPS.

3.3. Nitrogen Analysis

Figure 7 shows the CS percentages calculated from the amount of nitrogen determined by the Kjeldahl method. Since the original PLA fiber contains nitrogen, the CS amounts of the CS-coated PLA fibers were calculated after the subtraction of this nitrogen content, and were determined to be 0.33%, 0.38%, 0.58%, 0.73%, and 1.08% for C1–C5, respectively. C1, with only one CS layer, displayed a larger step in CS content than those of the other samples because C1 was formed by direct CS coating on the plasma-treated PLA surface. These results suggest that plasma treatment facilitates the efficient coating of the PLA surface by CS. The amount of CS was observed to increase proportionately with the number of CS-coating layers; this also agrees with similar results for CS/Alginate multilayer films obtained by Gabriela et al. [32]. In addition, C5 contained over 1.0 wt% CS.

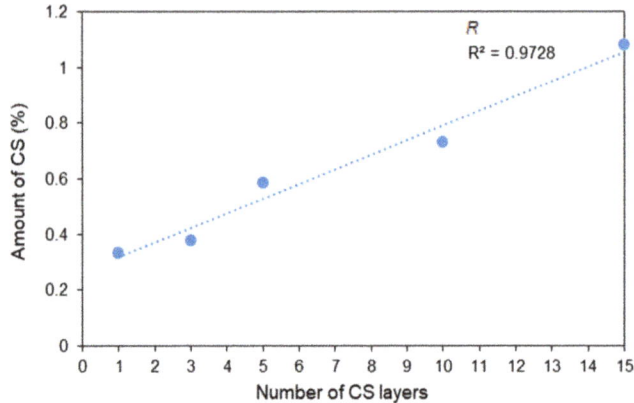

Figure 7. Amount of CS, calculated from the Kjeldahl-determined amount of nitrogen, as a function of the number of CS layers.

Table 2 lists the thicknesses of the CS layers calculated from the Kjeldahl-determined data. The coating thicknesses of C1–C5 were found to be 31.5, 35.9, 55.8, 69.7 and 103.7 nm, respectively. Clearly, the CS and SA layers are very thin and uniform, and do not affect the tensile properties of the fibers; all samples exhibit similar morphologies by SEM. These results confirm that the LBL method improves the amount of CS on the fibers.

3.4. Degradation Testing

Figures 8–10 display the rates of weight change, SEM images, and tensile strengths, respectively, of P1, P3, C3, and C5. Figures 8 and 9 reveal that all degraded samples exhibited at least 95% of their original weights and retained their shapes after five months of immersion in PBS solution. The observed decreases in weight indicate that the PLA samples become hydrolyzed in PBS over time. No samples could be retrieved from the PBS solution when immersed for longer than six months because the PLA fibers had collapsed by hydrolysis. The CS-coated fibers appeared to be slightly more stable than the uncoated plasma-treated fibers, because the polyelectrolyte layers of CS and SA had low solubility. However, the difference in stability was not significant; thus, it was concluded that the degradability of the fiber coated with CS and SA was maintained.

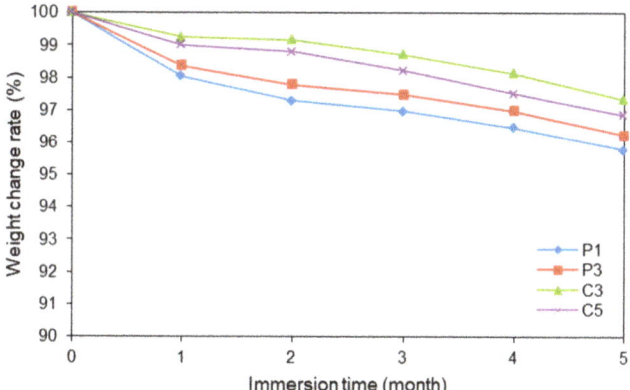

Figure 8. Rates of weight change of various fibers in PBS.

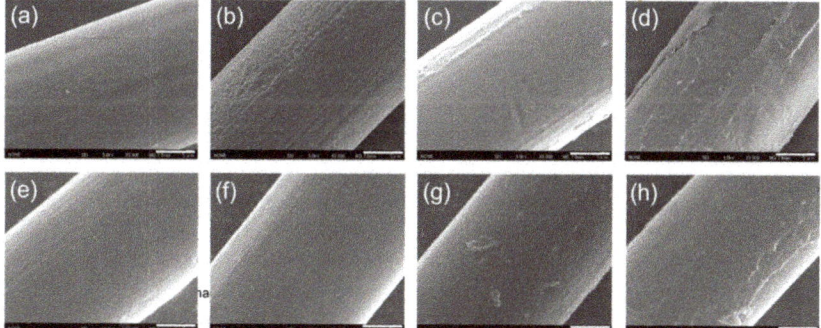

Figure 9. SEM images of PLA fibers: (**a**) P1; (**b**) P3; (**c**) C3; and (**d**) C5 before degradation testing; and (**e**) P1; (**f**) P3; (**g**) C3; and (**h**) C5 after six months in PBS (magnification: 5000×, scale bar: 5 µm).

The SEM images reveal that the PLA surfaces, as well as the fiber diameters, changed little after six months of immersion in PBS solution. We suggest that the observed weight losses are due to internal fiber hydrolysis. The tensile strengths of the degraded samples at three and four months were 90% and 80% of those of the original PLA fiber, respectively. The rates of weight change of the samples decreased enormously over five months, and the samples degraded for six months were too weak to subject to tensile testing; however, their surface morphologies did not change as extensively as their tensile properties. We suggest that the PLA molecules reduce their molecular weights by hydrolysis. Moreover, the degradation behavior of all samples was similar. The existence or not of a coating on the PLA fiber did not appear to influence the degradation of the fiber, which is ascribable to the extremely thin coating, as evidenced by nitrogen analysis and SEM. These results confirm that these PLA fibers are resistant to PBS for four months and degrade over long times.

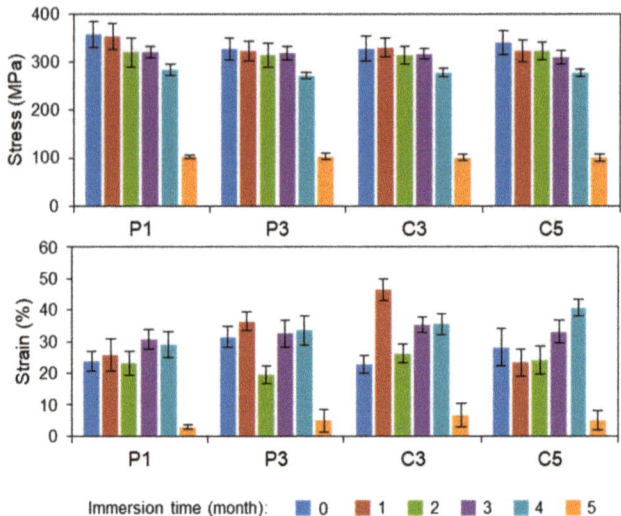

Figure 10. Tensile properties of the degraded samples.

4. Conclusions

Novel CS-coated PLA fibers were prepared by plasma treatment and the LBL method. SEM, tensile testing, and XPS reveal that plasma treatment facilitates the damage-free and effective coating of PLA with CS. In particular, PLA fibers treated with plasma at 100 W for 60 s exhibit almost identical properties to those of the original PLA fibers. Nitrogen analysis revealed that the LBL method effectively increased the amount of coated CS; moreover, the mechanical strengths of the obtained fibers were maintained following coating with CS. These PLA fibers were resistant to PBS for four months during degradation testing. In addition, these samples exhibited similar degradation behavior; PLA hydrolysis was unaffected by the presence or absence of the extremely thin ionic polymer. The presence of CS on PLA fiber is expected to improve the surface property of the fiber, owing to the antibacterial and wound-healing abilities of CS. Therefore, these CS-coated PLA fibers are suitable for use in biomaterials applications as suture threads.

Author Contributions: T.F. and H.T. designed experiments. D.K. and R.I. performed the experiments. D.K. wrote the manuscript.

Acknowledgments: This work was financially supported by Private University Research Branding Project, MEXT, 2016–2020 and in part by the Kansai University Outlay Support for Establishing Research Centers, 2016−2017. "Development and application of biocompatible polymer materials having sol-gel transition".

Conflicts of Interest: The authors declare no conflict of interest.

References

1. Tomita, K.; Kuroki, Y.; Nagai, K. Isolation of thermophiles degrading poly(L-lactic acid). *J. Biosci. Bioeng.* **1999**, *87*, 752–755. [CrossRef]
2. Ikada, Y.; Tsuji, H. Biodegradable polyesters for medical and ecological applications. *Macromol. Rapid Commun.* **2000**, *21*, 117–132. [CrossRef]
3. Tsuji, H.; Ikada, Y. Blends of aliphatic polyesters. II. Hydrolysis of solution-cast blends from poly(L-lactide) and poly(ε-caprolactone) in phosphate-buffered solution. *J. Appl. Polym. Sci.* **1998**, *67*, 405–415. [CrossRef]
4. Lunt, J. Large-scale production, properties and commercial applications of polylactic acid polymers. *Polym. Degrad. Stabil.* **1998**, *59*, 145–152. [CrossRef]
5. Saravanan, M.; Domb, A.J. A contemporary review on—Polymer stereocomplexes and its biomedical application. *Eur. J. Nanomed.* **2013**, *5*, 81–96. [CrossRef]

6. Jain, R.A. The manufacturing techniques of various drug loaded biodegradable poly(lactide-*co*-glycolide) (PLGA) devices. *Biomaterials* **2000**, *21*, 2475–2490. [CrossRef]
7. Mikos, A.G.; Lyman, M.D.; Freed, L.E.; Langer, R. Wetting of poly(L-lactic acid) and poly(DL-lactic-co-glycolic acid) foams for tissue culture. *Biomaterials* **1994**, *15*, 55–58. [CrossRef]
8. Nagahama, K.; Mori, Y.; Ohya, Y.; Ouchi, T. Biodegradable nanogel formation of polylactide-grafted dextran copolymer in dilute aqueous solution and enhancement of its stability by stereocomplexation. *Biomacromolecules* **2007**, *8*, 2135–2141. [CrossRef] [PubMed]
9. Rasal, R.M.; Janorkar, A.V.; Hirt, D.E. Poly(lactic acid) modifications. *Prog. Polym. Sci.* **2010**, *35*, 338–356. [CrossRef]
10. Furuike, T.; Nagahama, H.; Chaochai, T.; Tamura, H. Preparation and characterization of chitosan-coated poly(L-lactic acid) fibers and their braided rope. *Fibers* **2015**, *3*, 380–393. [CrossRef]
11. Zhu, A.; Zhang, M.; Wu, J.; Shen, J. Covalent immobilization of chitosan/heparin complex with a photosensitive hetero-bifunctional crosslinking reagent on PLA surface. *Biomaterials* **2002**, *23*, 4657–4665. [CrossRef]
12. Ding, Z.; Chen, J.; Gao, S.; Chang, J.; Zhang, J.; Kang, E.T. Immobilization of chitosan onto poly-L-lactic acid film surface by plasma graft polymerization to control the morphology of fibroblast and liver cells. *Biomaterials* **2004**, *25*, 1059–1067. [CrossRef]
13. Tsuji, H.; Ishida, T. Poly(L-lactide). X. Enhanced surface hydrophilicity and chain-scission mechanisms of poly(L-lactide) film in enzymatic, alkali, and phosphate-buffered solutions. *J. Appl. Polym. Sci.* **2003**, *87*, 1628–1633. [CrossRef]
14. Park, G.E.; Pattison, M.A.; Park, K.; Webster, T.J. Accelerated chondrocyte functions on NaOH-treated PLGA scaffolds. *Biomaterials* **2005**, *26*, 3075–3082. [CrossRef] [PubMed]
15. Nagahama, H.; Nwe, N.; Jayakumar, R.; Koiwa, S.; Furuike, T.; Tamura, H. Novel biodegradable chitin membranes for tissue engineering applications. *Carbohyd. Polym.* **2008**, *73*, 295–302. [CrossRef]
16. Tamura, H.; Nagahama, H.; Tokura, S. Preparation of chitin hydrogel under mild conditions. *Cellulose* **2006**, *13*, 357–364. [CrossRef]
17. Jayakumar, R.; Menon, D.; Manzoor, K.; Nair, S.V.; Tamura, H. Biomedical applications of chitin and chitosan based nanomaterials—A short Review. *Carbohyd. Polym.* **2010**, *82*, 227–232. [CrossRef]
18. Anitha, A.; Rani, V.V.D.; Krishna, R.; Sreeja, V.; Selvamurugan, N.; Nair, S.V.; Tamura, H.; Jayakumar, R. Synthesis, characterization, cytotoxicity and antibacterial studies of chitosan, O-carboxymethyl and N,O-carboxymethyl chitosan nanoparticles. *Carbohyd. Polym.* **2009**, *78*, 672–677. [CrossRef]
19. Carvalho, C.R.; López-Cebral, R.; Silva-Correia, J.; Silva, J.M.; Mano, J.F.; Silva, T.H.; Freier, T.; Reis, R.L.; Oliveira, J.M. Investigation of cell adhesion in chitosan membranes for peripheral nerve regeneration. *Mater. Sci. Eng. C* **2017**, *71*, 1122–1134. [CrossRef] [PubMed]
20. Dodane, V.; Vilivalam, V.D. Pharmaceutical applications of chitosan. *Pharm. Sci. Technol. Today* **1998**, *1*, 246–253. [CrossRef]
21. Jayakumar, R.; Prabaharan, M.; Kumar, P.T.S.; Nair, S.V.; Tamura, H. Biomaterials based on chitin and chitosan in wound dressing applications. *Biotechnol. Adv.* **2011**, *29*, 322–337. [CrossRef] [PubMed]
22. Chen, Y.; Javvaji, V.; MacIntire, I.C.; Raghavan, S.R. Gelation of vesicles and nanoparticles using water-soluble hydrophobically modified chitosan. *Langmuir* **2013**, *29*, 15302–15308. [CrossRef] [PubMed]
23. Davis, T.A.; Llanes, F.; Volesky, B.; Diaz-Pulido, G.; McCook, L.; Mucci, A. ^{1}H–NMR study of Na alginates extracted from sargassum spp. in relation to metal biosorption. *Appl. Biochem. Biotech.* **2003**, *110*, 75–90. [CrossRef]
24. Gomez, C.G.; Lambrecht, M.V.P.; Lozano, J.E.; Rinaudo, M.; Villar, M.A. Influence of the extraction-purification conditions on final properties of alginates obtained from brown algae (*Macrocystis pyrifera*). *Int. J. Biol. Macromol.* **2009**, *44*, 365–371. [CrossRef] [PubMed]
25. Tamura, H.; Tsuruta, Y.; Tokura, S. Preparation of chitosan-coated alginate filament. *Mater. Sci. Eng. C* **2002**, *20*, 143–147. [CrossRef]
26. Lee, K.Y.; Mooney, D.J. Alginate: Properties and biomedical applications. *Prog. Polym. Sci.* **2012**, *37*, 106–126. [CrossRef] [PubMed]
27. Zhang, M.; Lin, H.; Shen, L.; Liao, B.-Q.; Wu, X.; Li, R. Effect of calcium ions on fouling properties of alginate solution and its mechanisms. *J. Membr. Sci.* **2017**, *525*, 320–329. [CrossRef]

28. Ariga, K.; Hill, J.P.; Ji, Q. Layer-by-layer assembly as a versatile bottom-up nanofabrication technique for exploratory research and realistic application. *Phys. Chem. Chem. Phys.* **2007**, *9*, 2319–2340. [CrossRef] [PubMed]
29. Antipov, A.A.; Sukhorukov, G.B.; Donath, E.; Mohwald, H. Sustained release properties of polyelectrolyte multilayer capsules. *J. Phys. Chem. B* **2001**, *105*, 2281–2284. [CrossRef]
30. Kyung, W.K.-H.; Kim, S.-H.; Siratori, S. Preparation and characterization of antithrombogenic chitosan/alginate films with enhanced physical stability by cross-linking using layer-bylayer. *MATEC Web Conf.* **2013**, *4*, 05008. [CrossRef]
31. Wan, Y.; Tu, C.; Yang, J.; Bei, J.; Wang, S. Influences of ammonia plasma treatment on modifying depth and degradation of poly(L-lactide) scaffolds. *Biomaterials* **2006**, *27*, 2699–2704. [CrossRef] [PubMed]
32. Martins, G.V.; Merino, E.G.; Mano, J.F.; Alves, N.M. Crosslink Effect and Albumin Adsorption onto Chitosan/Alginate Multilayered Systems: An in situ QCM-D Study. *Macromol. Biosci.* **2010**, *10*, 1444–1455. [CrossRef] [PubMed]

© 2018 by the authors. Licensee MDPI, Basel, Switzerland. This article is an open access article distributed under the terms and conditions of the Creative Commons Attribution (CC BY) license (http://creativecommons.org/licenses/by/4.0/).

Article

Applying Image Processing to the Textile Grading of Fleece Based on Pilling Assessment

Mei-Ling Huang * and Chien-Chang Fu

Department of Industrial Engineering & Management, National Chin-Yi University of Technology, Taichung 411, Taiwan; chienchangfu@outlook.com
* Correspondence: huangml@ncut.edu.tw; Tel.: +886-4-23924505-7653

Received: 3 August 2018; Accepted: 26 September 2018; Published: 28 September 2018

Abstract: Textile pilling causes an undesirable appearance on the surface of garments, which is a long-standing problem. In this study, textile grading of fleece based on pilling assessment was performed using image processing and machine learning methods. Two image processing methods were used. The first method involved using the discrete Fourier transform combined with Gaussian filtering, and the second method involved using the Daubechies wavelet. Furthermore, binarization was used to segment the textile pilling from the background. Morphological and topological image processing methods were applied to extract the essential characteristics of textile image information to establish a database for the textile. Finally, machine learning methods, namely the artificial neural network (ANN) and the support vector machine (SVM), were used to objectively solve the textile grading problem. When the Fourier-Gaussian method was used, the classification accuracies of the ANN and SVM were 96.6% and 95.3%, and the overall accuracies of the Daubechies wavelet were 96.3% and 90.9%, respectively.

Keywords: textile; pilling; image processing; machine learning

1. Introduction

Standard testing methods of textile grading are based on wear resistance numbers and visual determination of the textile grade. These visual detection methods may cause errors. Using high-resolution images, Zheng et al. [1] proposed a two-step method to detect the yarn location and weave structure in a woven fabric. The results indicated that their methodology was effective in detecting the fabric structure and yarn float. Liu et al. [2] proposed an algorithm to integrate local texture features and whole-image texture information for detecting texture defects. The techniques used by Liu et al. [2] included the local binary pattern, salient region detection, and segmentation.

In a related work by Kuo et al. [3], a textile image processing technique was applied to detect textile defects; the study applied the wavelet transform to extract textile features and classified textile images into seven categories by using a neural network. The proposed method saved grading time and enhanced inspection capability. Deng et al. [4] applied the continuous wavelet transform to the pilling of textile images. Six textile features, namely entropy ratio, volume, area, area standard deviation, height standard deviation, and the location deviation coefficient, were confirmed as variables in the input layer of the back-propagation neural network to classify textile grades. Bissi et al. [5] presented an algorithm using the Gabor filter and principal component analysis to detect texture defects on a patch basis. Uniformly textured fabrics, visibly textured fabrics, and grid-like structured fabrics were tested, and the results outperformed those of other relevant studies.

Saharkhiz et al. [6] applied the two-dimensional (2D) fast Fourier transform method for image processing, considering low-pass filtering and a suitable cutoff frequency. Three features were extracted: The number of pilling points, volume, and area. This study used clustering methods, including median-cut, k-means, and competitive learning, to classify textile pilling. Yun et al. [7] applied the fast Fourier transform and fast wavelet transform for image processing. The following three crucial features were identified: The number of pilling points, total pixel area of pilling, and the sum of the gray values of pilling images. Jing et al. [8] developed an objective fabric pilling evaluation method that incorporated the wavelet transform and local binary pattern. Their study normalized the values of extracted features and applied the support vector machine as a classifier to evaluate pilling grades; the results indicated a grading accuracy of 95%. Based on a standard of the Woolmark SM50 blanket set, Zhang et al. [9] proposed an objective pilling evaluation method for nonwoven fabrics. The developed method was a combination of pilling identification, characterization, and a neural network; the results indicated the proposed classifier to be feasible.

This paper proposes an objective grading method for fleece based on pilling assessment to replace the subjective visual grading method and make textile grading robust through hierarchical steps. First, textile images are processed using image processing methods. In image processing, the Gaussian filter is ideal for image smoothing, and the Daubechies wavelet is commonly used in the medical field. In this study, two image processing methods were used, and their results were compared. One method involved the discrete Fourier transform (DFT) combined with Gaussian filtering, whereas the other method involved Daubechies wavelet filtering. After filtering the images, binarization was used to segment the pilling from the background. The textile grade worsens as the pilling increases. To develop an objective grading system, morphological and topological image processing methods were used to extract the essential parameters of the pilling. The textile image database featured five parameters, namely the number of pilling points, pilling area, average pilling area, pilling area ratio, and pilling density, which were extracted for each textile image. Artificial neural network (ANN) and support vector machine (SVM) models were used to train the data and objectively classify the textile grade.

2. Experiment

The proposed method involved five steps: Data acquisition, image processing, feature extraction, model building, and performance evaluation, see Figure 1.

Step 1: A total of 320 representative samples were collected and classified as grade 2, 3, 4, or 5. Each grade comprised of 80 samples. A charge coupled device (CCD) camera was used to capture the fabric image for constructing the grayscale image dataset of the 320 samples. The camera was placed over the sample to capture the image.

Step 2: The obtained grayscale images were filtered using two methods: the DFT method combined with Gaussian filtering and the Daubechies wavelet method. The DFT method combined with Gaussian filtering was used to smooth the grayscale images. The Daubechies wavelet method was used to compresses the grayscale images. To retain the information of textile pilling, the filtered images were transformed into black and white through binarization.

Step 3: Morphological and topological image processing methods were used to extract the essential parameters of the pilling. Five parameters were considered in this study, namely the number of pilling points, pilling area, average pilling area, pilling area ratio, and pilling density. The size of the database was 320×5.

Step 4: ANN and SVM models were used to classify the pilling. Finally, the classification rates of different classifiers were compared.

2.1. Sample

Textiles used in this study complied with the ISO 12945-2:2000 Martindale wear standards of textile grading. Five grades are present in this standard. Textiles categorized as grade 1 have the lowest quality, whereas textiles categorized as grade 5 have the highest quality. No grade 1 textiles were used in this study. Thus, 80 samples in each grade between 2 and 5 were examined. The appearance of grades 2–5 is shown in Table 1.

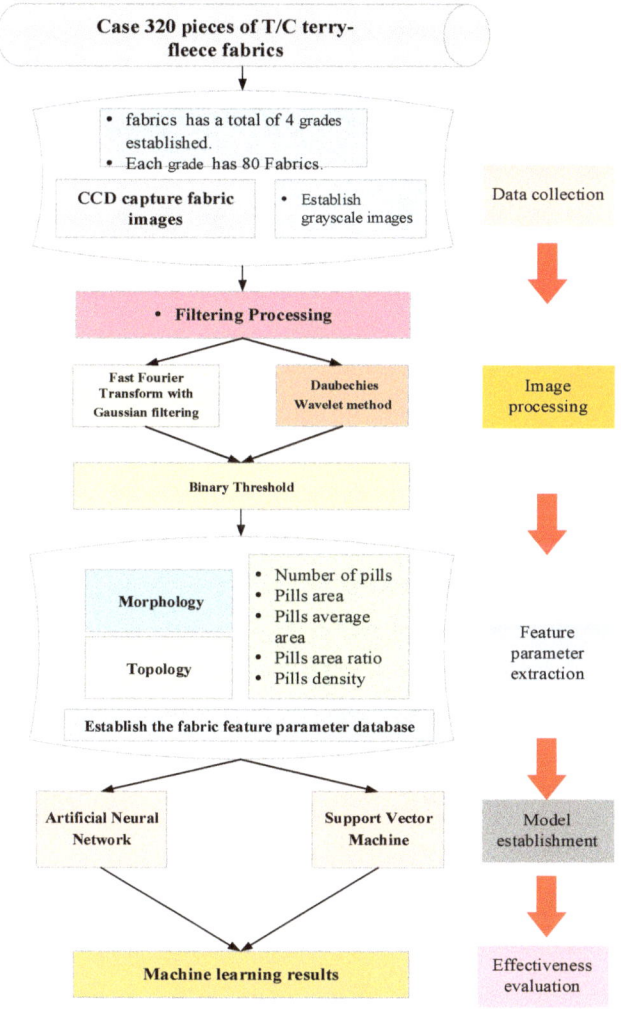

Figure 1. Research flowchart.

Table 1. Sample of the textile grade.

2.2. Textile Sample Collection

The specifications of the experimental equipment are listed in Table 2. The Basler A601f-2 CCD camera (Basler Vision Technologies, Hsinchu, Taiwan) with a resolution of 656 × 491 pixels and a frequency of 58 fps was used in this study. The fabric examined in this study was fleece; a testing sample was placed on the work platform with a 13 W light source, as displayed in Figure 2. The low-angle oblique illumination method suggested by Kayseri and Kirtay [10] and Saharkhiz et al. [6] was used to obtain positive lighting for capturing textile images. Low-angle oblique illumination is commonly used to detect and examine raised and depressed flaws on a flat surface. Low-angle oblique illumination reduces shadow generation and produces evenly illuminated images. Its disadvantage is that the flat surface may have a strong reflection. Table 3 presents the grayscale images of textile grades 2–5.

Table 2. Specifications of the experimental equipment.

CCD	Model	Pixels	Frequency
	Basler A601f-2	656 × 491	58FPS
Lens	**Model**	**Focal Length**	**Maximum Aperture Ratio**
	M2514-MP2	25 mm	1:1.4

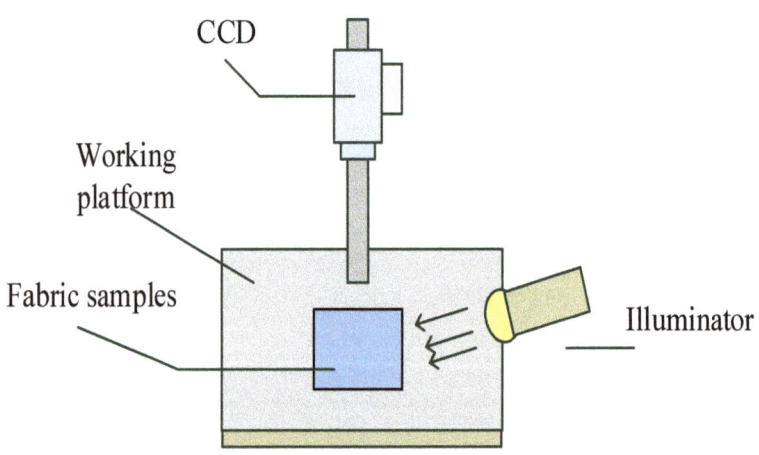

Figure 2. Image capture diagram.

Table 3. CCD images of textiles.

Picture	Grade 2	Grade 3	Grade 4	Grade 5
Grayscale image				

2.3. Image Processing

2.3.1. DFT with Gaussian Filtering

A low-pass filter is commonly used for image blurring or smoothing, whereas a high-pass filter is used for image edge enhancement. The Gaussian filter is the most widely used smoothing filter. The formula of the Gaussian filter is expressed as follows:

$$H(u,v) = \frac{1}{2\pi\sigma^2} e^{-(u^2+v^2)/2\sigma^2} \quad (1)$$

where u and v are the horizontal and vertical distances from the origin, respectively, and σ is the standard deviation of the Gaussian distribution. The DFT is designed for discrete and periodical signals. The DFT formula for discrete signals with $u = 0, 1, 2, \cdots, M-1$ and $v = 0, 1, 2, \cdots, N-1$ is presented in Equation (2). Using Equation (2), we can obtain an inverse conversion $[F(x,y)]$ image with the signal $x = 0, 1, 2, \cdots M-1$ and $y = 0, 1, 2, \cdots N-1$. The formula of the inverse DFT is provided in Equation (3).

$$F(u,v) = \sum_{u=0}^{M-1} \sum_{y=0}^{N-1} f(x,y) e^{-j2\pi(ux/M+vy/N)} \quad (2)$$

$$F(x,y) = \frac{1}{MN} \sum_{u=0}^{M-1} \sum_{y=0}^{N-1} f(u,v) e^{j2\pi(ux/M+vy/N)} \quad (3)$$

In this study, the DFT and Gaussian filter were combined for image processing. A textile image of size $M \times N$ is filtered using Equation (4).

$$g(x,y) = \Im^{-1}[H(u,v)F(u,v)] \quad (4)$$

The purpose of the filtering procedure is to dilute the background texture of the textile. The image filtering process is illustrated in Figure 3. Matlab software was used to perform the filtering procedure. In the conversion of Gaussian filtering, we must set a value for the parameter, sigma. The default sigma value in Matlab is 0.5 for an image of size 3 × 3 pixels. The size of the textile images used in this study was 320 × 240 pixels. Several values of sigma between 10 and 50 were tested to compare the filtering performance for original and filtered textile images, see Table 4. A small value of sigma results in blurry images, as shown in Table 4. The larger the sigma value, the clearer the filtered image is.

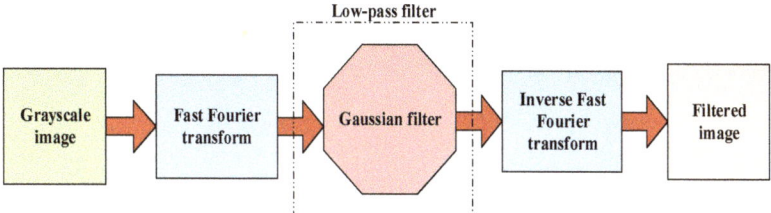

Figure 3. Filtering process.

Table 4. Filtered images with corresponding sigma values.

Original Image	Sigma Value	Filtered Image
	10	
	20	
	30	
	40	
	50	

The mask size of the filter was also considered in this study. The larger the value of sigma, the higher the distortion in the image is. When the sigma value was set to 50, the mask was larger than the image size, and the image exhibited distortion. Therefore, the sigma value of the filter mask size was set to 40 in this study. The filtered images and corresponding sigma values of the mask size are presented in Table 5.

Table 5. Filter mask size.

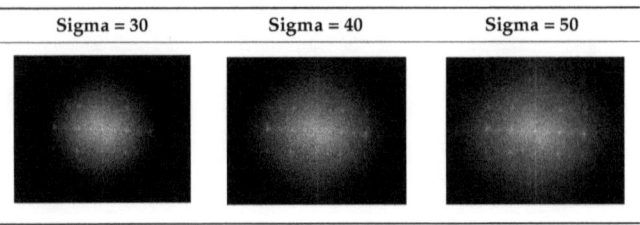

Sigma = 30	Sigma = 40	Sigma = 50

2.3.2. Daubechies Wavelet

The Daubechies wavelet comprises the scaling function $\varnothing(x)$ and wavelet function $\psi(x)$, as presented in Equations (5) and (6), respectively.

$$\varnothing(x) = \sum_{k=0}^{n-1} a_k \phi(2x - k) \qquad (5)$$

$$\psi(x) = \sum_{k=0}^{M-1} b_k \phi(2x - k) \qquad (6)$$

The commonly used filter lengths are between 2 and 20. Daubechies 4 (called D4) was used in this study. The expanded formulas for the Daubechies wavelet are as follows:

$$\varnothing(x) = h_0 \varnothing(2x) + h_1 \varnothing(2x - 1) + h_2 \varnothing(2x - 2) + h_1 \varnothing(2x - 2) \qquad (7)$$

$$\psi(x) = h_0 \varnothing(2x - 1) + h_1 \varnothing(2x) + h_2 \varnothing(2x + 1) + h_3 \varnothing(2x + 2) \qquad (8)$$

Commonly used image evaluation methods include the mean square error (MSE), signal to noise ratio (SNR), and peak signal to noise ratio (PSNR). Smaller MSE and larger SNR and PSNR values represent superior image quality. In this study, these indices were applied to evaluate image processing quality. In the Daubechies wavelet step, various scale values were simulated, and the evaluation indices MSE, SNR, and PSNR were computed to determine the image quality. Table 6 presents the image evaluation results for various scales. This study used Scale 1 because it had the lowest MSE value and the highest SNR and PSNR values.

Table 6. Image evaluation results.

Image	MSE	SNR	PSNR
Original	0	Infinite	Infinite
Scale 1	2.010	23.902	33.491
Scale 2	4.937	19.570	29.170
Scale 3	8.468	16.118	25.234
Scale 4	8.656	16.306	25.924
Scale 5	9.269	15.609	25.239
Scale 6	10.108	14.770	24.414
Scale 7	12.562	13.128	22.693

Based on the aforementioned discussion and the conversion steps of Umbaugh [11], six conversion steps were designed for the textile images in this study. Figure 4 shows the result of the wavelet transform.

1. Wavelet conversion step 1: Use the low-pass filter to compute the convolution in the horizontal direction of the image.
2. Wavelet conversion step 2: Use the low-pass filter to compute the convolution in the vertical direction of the image.
3. Wavelet conversion step 3: Use the high-pass filter to compute the convolution in the horizontal direction of the image.
4. Wavelet conversion step 4: Use the high-pass filter to compute the convolution in the vertical direction of the image.
5. Wavelet conversion step 5: Use the low-pass filter to compute the convolution in the vertical direction of the image.
6. Wavelet conversion step 6: Use the high-pass filter to compute the convolution in the vertical direction of the image.

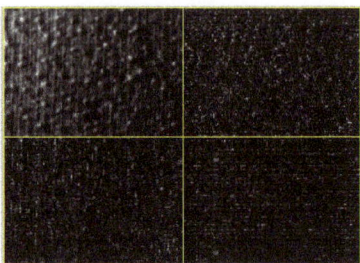

Figure 4. Result of the wavelet transform.

2.3.3. Binarization

Binarization of the textile image was performed using the im2bw function in the Matlab DIP toolbox to convert the grayscale image to black and white. Thus, the converted image included only pure black and pure white pixels with values of 0 and 255, respectively. The default threshold value was 0.5. Thus, if the pixel value of the image was higher than 127.5 (255 × 0.5), the final value was set to 255. The final value was set to 0 if the pixel value was less than 127. Table 7 lists the results for the threshold values of 0.4, 0.5, and 0.6. A value of 0.6 was used in this study because it could extract the most critical characteristics of pills.

Table 7. Results using different thresholds.

Threshold 0.4	Threshold 0.5	Threshold 0.6

2.4. Step 3: Feature Extraction of the Filtered Textile Image

2.4.1. Morphological Image Processing

Morphological image processing was used to extract the essential characteristics of the filtered textile images. Erosion and dilation are the two basic operations in morphological image processing. The "opening operation" was used to handle binary images.

Opening operation: The given A and structuring element B (denoted by $A\ B$) represent an erosion followed by a dilation.

$$AB = (A \ominus B) \oplus B \qquad (9)$$

In the opening operation, erosion shrinks the image to remove any unessential information. The regions remaining after shrinking are dilated to the original size to highlight the essential pilling characteristics of the textile image.

2.4.2. Image Topology

Image topology involves analyzing discrete objects in a 2D digital image and constructing a mathematical model to considerably simplify the algorithmic design. Assume that A is the current pixel. The top neighbor of A is o, and the left neighbor of A is e. A four-adjacency relationship defined for pixel A by adjacent positions is displayed in Figures 5 and 6.

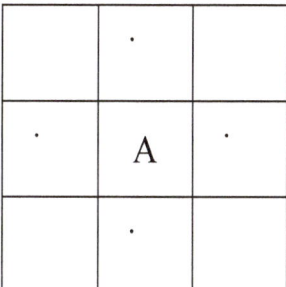

Figure 5. Four neighbors.

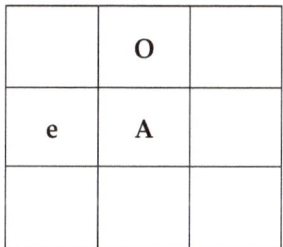

Figure 6. Four adjacency positions.

In this study, the topological approach was used, where pixels in the image were defined as foreground pixels and pixels not in the image were defined as background pixels, to calculate the number of pilling points and pilling area from the textile image.

2.4.3. Database Creation

The commonly used characteristics for textile grading are described as follows. Saharkhiz et al. [6] used the number of pilling points, pilling volume, and pilling area for textile grading. Eessouki et al. [12] used the number of pilling points, average pilling area, pilling area ratio, and pilling density extracted from the textile image for textile grading. The grading method proposed in this study involved examining five characteristics, namely the number of pilling points, pilling area, average pilling area, pilling area ratio, and pilling density extracted from the textile image. These five characteristics were examined for each sample. Based on the number of pilling points and the pilling area obtained from the textile image, the average pilling area, pilling area ratio, and pilling density were computed using Equations (10)–(12). The average pilling area is the pilling area divided by the number of pilling points. The pilling area ratio is the pilling area divided by the image size. The pilling density is the pilling area divided by the number of pilling points.

$$\text{Average pilling area} = \frac{\text{Pilling area}}{\text{Number of pilling points}} \quad (10)$$

$$\text{Pilling area ratio} = \frac{\text{Pilling area}}{\text{Image size}} \quad (11)$$

$$\text{Pilling density} = \frac{\text{Number of pilling points}}{\text{Image size}} \quad (12)$$

2.5. Step 4: Model Building for Textile Grading

2.5.1. ANN

An ANN is a machine learning tool in artificial intelligence that mimics the thoughts of humans. It can be applied in a wide range of areas, such as nonlinear models. An ANN usually includes three layers: input, hidden, and output layers. The presence of a large number of neurons in the hidden layer increases the accuracy of the results; however, it slows the convergence rate of the network. The data from Section 2.4.3 were used as the input. The number of neurons used in this study was the average number of variables in the input and output layers [i.e., (5 + 1)/2 = 3]. The number of pilling points, pilling area, average pilling area, pilling area ratio, and pilling density extracted from the textile image were used as input items in the ANN. The training set comprised 80% of the database, and the remaining 20% was used as the test set. Ten-fold cross-validation was applied in the ANN.

2.5.2. SVM

SVMs use a hyperplane in n-dimensional space to distinguish between two or more categories of information. The linear, polynomial, and radial functions are the commonly used core functions in an SVM. The performance of the radial function is superior to that of the other functions for high-dimensional information. Thus, the radial function was used in this study for textile grading. The Weka software (Version 3.6, The WEKA Workbench, Machine Learning Group at the University of Waikato, New Zealand) was used for SVM execution. The input items in the SVM were the same as those in the ANN. The parameter settings in the SVM are described as follows. The sigma value was set to 0.05, and the penalty function (C value) was set to 1 by default. The SVM termination condition was set to 1.0×10^{-6}. The training set comprised 80% of the database, and the remaining 20% was used as the test set. Ten-fold cross-validation was applied in the SVM.

3. Experiment Results

3.1. Image Processing Results

Image processing results are presented in this section. Table 8 presents the image processing results of grade 2 textiles for each step. The second row in Table 8 represents the original textile image. The filtering results obtained using the DFT method combined with Gaussian filtering and the Daubechies wavelet method are displayed in the third row of Table 6. The textile image filtered using the DFT method combined with Gaussian filtering exhibited a diluted background texture. The Daubechies wavelet method filtered and compressed the original textile image to remove noise and enhance pilling characteristics. Through binarization, the image background was separated from the pilling. However, some white spots remained in the image. Morphological methods were then used to retain essential information and remove unessential information. The results obtained through binarization and morphological methods are displayed in the fourth and fifth rows in Table 6, respectively. The pilling features were highlighted and extracted to construct the dataset for machine learning to complete the final step of textile grading.

Table 8. Image processing results of the textile image.

Grade 2	Fourier with Gaussian	Daubechies Wavelet
The original image		
Filtering processing		
Binarization		
Morphology		
Mark the results		

3.2. Classification Results Using Machine Learning

Using the dataset from Section 2.4.3, ANN and SVM models were executed. The grading performance of the models was compared. When the Fourier-Gaussian method was used, the classification accuracies of the ANN and SVM were 96.6% and 95.3%, respectively, see Tables 9 and 10. Compared with human inspection and grading, textile grading through machine learning requires less processing time and achieves higher classification accuracy.

Table 9. Results of Fourier-Gaussian Method with artificial neural network (ANN).

Predict Actual	Grade 2	Grade 3	Grade 4	Grade 5		
Grade 2	74	5	1	0	Correct number	309
Grade 3	5	75	0	0	Wrong number	11
Grade 4	0	0	80	0		
Grade 5	0	0	0	80	Accuracy	96.6%

Table 10. Results of Fourier-Gaussian Method with support vector machine (SVM).

Predict Actual	Grade 2	Grade 3	Grade 4	Grade 5		
Grade 2	80	0	0	0	Correct number	305
Grade 3	12	68	0	0	Wrong number	15
Grade 4	0	3	77	0		
Grade 5	0	0	0	80	Accuracy	95.3%

In the experiment, each grade comprised of 80 samples. When the Fourier-Gaussian method was used with the ANN, five grade 2 samples were classified as grade 3, one grade 2 sample was classified as grade 4, and five grade 3 samples were classified as grade 2, see Table 9. All the samples of grades 4 and 5 were correctly classified. Of the 320 samples, 309 were correctly graded, and 11 were misgraded. Thus, the overall accuracy was 96.6%.

When the Fourier-Gaussian method was used with the SVM, 12 grade 3 samples were classified as grade 2, and three grade 4 samples were classified as grade 3, see Table 10. All the grade 2 and 5 samples were correctly classified. Of the 320 samples, 305 were correctly graded, and 15 were misgraded. Thus, the overall accuracy was 95.3%.

When the Daubechies wavelet method was used with the ANN, five grade 2 samples were classified as grade 3, three grade 2 samples were classified as grade 4, two grade 3 samples were classified as grade 2, and two grade 3 samples were classified as grade 4, see Table 11. All the samples of grades 4 and 5 were correctly classified. Of the 320 samples, 308 were correctly graded, and 12 were misgraded. Thus, the overall accuracy was 96.3%.

Table 11. Results of the Daubechies wavelet method with the ANN.

Predict Actual	Grade 2	Grade 3	Grade 4	Grade 5		
Grade 2	72	5	3	0	Correct number	308
Grade 3	2	76	2	0	Wrong number	12
Grade 4	0	0	80	0		
Grade 5	0	0	0	80	Accuracy	96.3%

When the Daubechies wavelet method was used with the SVM, 13 grade 2 samples were classified as grade 3, two grade 2 samples were classified as grade 4, 13 grade 3 samples were classified as grade 2, and one grade 3 sample was classified as grade 4, see Table 12. All the samples of grades 4 and 5 were correctly classified. Of the 320 samples, 291 were correctly graded, and 29 samples were misgraded. Thus, the overall accuracy was 90.9%.

Table 12. Results of the Daubechies wavelet method with the SVM.

Predict Actual	Grade 2	Grade 3	Grade 4	Grade 5		
Grade 2	65	13	2	0	Correct number	291
Grade 3	13	66	1	0	Wrong number	29
Grade 4	0	0	80	0		
Grade 5	0	0	0	80	Accuracy	90.9%

Table 13 summarizes the performance of the two filters for the same grade 2 sample. The number of pilling points obtained through the Fourier-Gaussian method was 61, which was higher than that obtained through the Daubechies wavelet method. With the Fourier-Gaussian method, the pilling area was 1552, whereas the pilling area with the Daubechies wavelet method was only 545. Thus, the performance of the Fourier-Gaussian filter was superior to that of the Daubechies wavelet filter.

Table 13. Comparison of Grade 2 Textile Filtering Methods.

Filtering Method	The Number of Pilling	Pilling Area
Fourier-Gaussian	61	1552
Daubechies wavelet	41	545

3.3. Comparison with Related Studies

The method used by Saharkhiz et al. [6] was applied to the textile images used in this study. The image processing results obtained using the Fourier transform with an ideal filter are presented in Table 14. After filtering, we followed the procedure described in Section 2.4. The final grading performance is listed in Table 15. When using the ANN, the grading accuracy of the Fourier transform with an ideal filter was 96.8%. When using the SVM, the grading accuracy of the Fourier transform with an ideal filter was 92.2%. When the ANN was selected as the classifier, the grading accuracies of the three image processing methods were similar. However, when the SVM was used, the DFT method with Gaussian filtering had the highest grading accuracy of 95.3%.

Table 14. Results of the Fourier transform with ideal filtering.

Method	Grade 2	Grade 3	Grade 4	Grade 5
Fourier-Ideal				

Table 15. Comparison of textile grading methods.

Author	Filter	ANN	SVM
Saharkhiz et al. [6]	Fourier-Ideal	96.8%	92.2%
This research	Fourier-Gaussian	96.6%	95.3%
	Daubechies wavelet	96.3%	90.9%

4. Discussion and Conclusions

Several machine learning and filtering methods have been proposed for pilling assessment. In the present study, a combination of the Fourier-Gaussian method with an ANN, a combination of Fourier-Gaussian with an SVM, a combination of Daubechies wavelet with an ANN, and a combination of Daubechies wavelet with an SVM were investigated; these have not been applied to the fabric grading of fleece in any related studies. To demonstrate the performance achieved by the proposed techniques, the method used by Saharkhiz et al. [6] was applied to textile images. The results indicated the grading accuracies to be close (approximately 96%) when ANN machine learning was used along with various filtering methods between the proposed methods and the related work. In addition, when using the SVM machine learning method, the grading accuracy from Fourier-Gaussian dominated the other two methods.

In this study, a textile grading method that comprises of image processing and machine learning classification is proposed as a replacement for visual textile grading. Two filtering methods were used in this study, namely the DFT combined with Gaussian filtering as well as Daubechies wavelet filtering. Binarization was used to segment the pilling from the background, and morphological and topological image processing methods were used to extract the essential parameters of the pilling. Two machine learning methods, ANN and SVM, were used to train the data and objectively classify the textile grade.

Although the grading accuracies of the proposed methods were satisfactory, other image filtering and machine learning methods can be used for textile grading in future research.

This study demonstrates various filtering techniques with different machine learning algorithms for the fabric grading of fleece. The effectiveness of the grading results depends on different colors and materials as well as other factors. Due to the diverse characteristics of fabric textures, colors, and densities, a general grading system for all types of fabric may not result in accurate grading; therefore, extracting crucial attributes and developing specific grading methods for different fabrics are the optimal solutions.

Author Contributions: M.-L.H. conceived of the presented idea and verified the analytical methods. C.-C.F. carried out the experiments. Mei-Ling Huang wrote the paper. All authors discussed the results and contributed to the final version of the manuscript.

Funding: This study was partially supported by the Ministry of Science and Technology, Taiwan, ROC, under the grant MOST 106-2221-E-167-015. The APC was funded by the Ministry of Science and Technology, Taiwan, ROC.

Conflicts of Interest: The authors declare no conflicts of interest.

References

1. Zheng, D.; Han, Y.; Hu, J.L. A new method for classification of woven structure for yarn-dyed fabric. *Text. Res. J.* **2014**, *84*, 78–95. [CrossRef]
2. Liu, Z.; Li, C.; Zhao, Q.; Liao, L.; Dong, Y. A fabric defect detection algorithm via context-based local texture saliency analysis. *Int. J. Cloth. Sci. Technol.* **2015**, *27*, 738–750. [CrossRef]
3. Jeffrey Kuo, C.-F.; Shih, C.-Y.; Huang, C.-C.; Wen, Y.-M. Image inspection of knitted fabric defects using wavelet packets. *Text. Res. J.* **2016**, *86*, 553–560. [CrossRef]
4. Deng, Z.; Wang, L.; Wang, X. An integrated method of feature extraction and objective evaluation of fabric pilling. *J. Text. Inst.* **2011**, *102*, 1–13. [CrossRef]
5. Bissi, L.; Baruffa, G.; Placidi, P.; Ricci, E.; Scorzoni, A.; Valigi, P. Automated defect detection in uniform and structured fabrics using gabor filters and pca. *J. Vis. Commun. Image Represent.* **2013**, *24*, 838–845. [CrossRef]
6. Saharkhiz, S.; Abdorazaghi, M. The Performance of Different Clustering Methods in the Objective Assessment of Fabric Pilling. *J. Eng. Fibers Fabr.* **2012**, *7*, 35–41.
7. Yun, S.Y.; Kim, S.; Park, C.K. Development of an objective fabric pilling evaluation method. II. Fabric pilling grading using artificial neural network. *Fibers Polym.* **2013**, *14*, 2157–2162. [CrossRef]
8. Jing, J.; Zhang, Z.; Kang, X.; Jia, J. Objective evaluation of fabric pilling based on wavelet transform and the local binary pattern. *Text. Res. J.* **2012**, *82*, 1880–1887. [CrossRef]
9. Zhang, J.; Wang, X.; Palmer, S. Objective pilling evaluation of wool fabrics. *Text. Res. J.* **2007**, *77*, 929–936. [CrossRef]
10. Kayseri, G.O.; Kirtay, E. Part 1. Predicting the Pilling Tendency of the Cotton Interlock Knitted Fabrics by Regression Analysis. *J. Eng. Fibers Fabr.* **2015**, *10*, 110–120.
11. Umbaugh, S.E. *Computer Vision and Image Processing: A Practical Approach Using CVIP Tools*; Prentice Hall: Bergen county, NJ, USA, 1998.
12. Eessouki, M.; Bukhari, H.A.; Hassan, M.; Qashqari, K. Integrated Computer Vision and Soft Computing System for Classifying the Pilling Resistance of Knitted Fabrics. *Fibres Text. East. Eur.* **2014**, *22*, 106–112.

© 2018 by the authors. Licensee MDPI, Basel, Switzerland. This article is an open access article distributed under the terms and conditions of the Creative Commons Attribution (CC BY) license (http://creativecommons.org/licenses/by/4.0/).

Article

A Time-Efficient Dip Coating Technique for the Deposition of Microgels onto the Optical Fiber Tip

Lorenzo Scherino [1], Martino Giaquinto [1], Alberto Micco [1], Anna Aliberti [1], Eugenia Bobeico [2], Vera La Ferrara [2], Menotti Ruvo [3], Armando Ricciardi [1,*] and Andrea Cusano [1,*]

[1] Optoelectronics Group, Department of Engineering, University of Sannio, 82100 Benevento, Italy; lorenzo.scherino@unisannio.it (L.S.); martino.giaquinto@unisannio.it (M.G.); alberto.micco@unisannio.it (A.M.); anna.aliberti@unisannio.it (A.A.)
[2] ENEA, Portici Research Center, P.le E. Fermi 1, 80055 Napoli, Italy; eugenia.bobeico@enea.it (E.B.); vera.laferrara@enea.it (V.L.F.)
[3] Institute of Biostructure and Bioimaging, National Research Council, 80143 Napoli, Italy; menotti.ruvo@unina.it
* Correspondence: aricciardi@unisannio.it (A.R.); a.cusano@unisannio.it (A.C.); Tel.: +39-0824-305601 (A.R.); +39-0824-305846 (A.C.)

Received: 1 August 2018; Accepted: 23 September 2018; Published: 28 September 2018

Abstract: The combination of responsive microgels and Lab-on-Fiber devices represents a valuable technological tool for developing advanced optrodes, especially useful for biomedical applications. Recently, we have reported on a fabrication method, based on the dip coating technique, for creating a microgels monolayer in a controlled fashion onto the fiber tip. In the wake of these results, with a view towards industrial applications, here we carefully analyze, by means of both morphological and optical characterizations, the effect of each fabrication step (fiber dipping, rinsing, and drying) on the microgels film properties. Interestingly, we demonstrate that it is possible to significantly reduce the duration (from 960 min to 31 min) and the complexity of the fabrication procedure, without compromising the quality of the microgels film at all. Repeatability studies are carried out to confirm the validity of the optimized deposition procedure. Moreover, the new procedure is successfully applied to different kinds of substrates (patterned gold and bare optical fiber glass), demonstrating the generality of our findings. Overall, the results presented in this work offer the possibility to improve of a factor ~30 the fabrication throughput of microgels-assisted optical fiber probes, thus enabling their possible exploitation in industrial applications.

Keywords: optical fiber sensors; lab-on-fiber technology; microgel; dip coating technique

1. Introduction

Integrating onto the optical fiber tip resonant nanostructures able to trap light at specific wavelengths is at the base of the Lab-on-Fiber (LOF) technology [1–6]. The resulting electromagnetic field confinement at sub-wavelength scale strongly enhances the light matter interaction, making possible to detect environmental changes onto the fiber surface as resonance wavelength shifts and/or intensity variation of the optical signal coupled to the fiber [4,5]. In this manner, ultra-sensitive optrodes based on LOF technology have been developed and exploited in biochemical applications, i.e., for detecting the presence of nano-sized bio-coating film resulting from molecular interactions. Moreover, as the optical fiber is, by nature, compatible with medical needles or catheters, this class of devices lends itself to in vivo detection [7].

The detection performances (such as limit of detection and response time) of LOF devices can be improved by exploiting functional materials used in combination with the resonant nanostructures. In this framework, we have recently proposed a LOF optrode integrated with microgels (MGs), i.e., 'smart' or responsive polymers, that change their size following environment changes [8–12]. Specifically, MGs are colloidal microsized hydrogel particles, which are synthesized to be selectively sensitive to physical (temperature) or chemical (i.e., pH variations, molecular binding events) parameters of interest when immersed in a liquid environment [13,14]. In biochemical applications, functionalized MGs offer the possibility to increase the target analyte loading capacity, and amplify the transduction of the optical signals [15–17]. In fact, in the case of MG assisted LOF optrodes, the resonance wavelengths are modulated by the MGs swelling/shrinking induced by the molecular binding, and not by the molecular binding itself occurring at the fiber surface [8].

In a previous work, we have reported a reliable fabrication strategy, based on the dip coating technique, for realizing, in a controlled fashion, monolayers of MGs onto the optical fiber tip [11]. The formation of the MGs film was controlled by selecting the parameters of the MGs solutions used during the dipping procedure. In particular, by setting low operating pH (3), low temperature (10 °C) and high concentration (5%) of particle dispersions, we achieved high coverage factor (CF) films. We have demonstrated that CFs larger than 90% warranted the maximum degree of light-MGs interactions onto the fiber tip, and thus the maximum responsivity to MGs swelling/collapsing induced by the specific external stimulus of interest [11,12].

However, the dip coating procedure used was very time consuming (about 15 h, excepting the dipping step) and it required the use of controlled temperature environment (oven) for the drying phase. In fact, the procedure started by dipping the fiber into the MGs dispersion for 1 h. Once the fiber probe was extracted, it was dried at a controlled temperature (45 °C) into an oven for 1 h. Then the probe was immersed in a deionized water bath for 12 h at room temperature under magnetic stirring in order to ensure breaking up of a possible multilayer. Finally, the probe was dried into an oven at 30 °C for 2 h. Although more fibers can be in principle deposited in parallel at the same time, the use of such time consuming process strongly limits the large scale production of LOF optrodes assisted by MGs.

With the aim of overcoming this limitation and improving the fabrication throughput, here we carefully analyze the influence of all the steps (dipping, rinsing, and drying) involved in the deposition procedure. Interestingly, we demonstrate that it is possible to significantly reduce the fabrication time while maintaining the same quality standard of the MGs films created onto the fiber tip surface. Specifically, we found that the MGs assisted optical fiber probe production throughput can be enhanced up to a factor ~30, without resorting to controlled temperature instruments during the drying phase. It is important to point out that optical monitoring of the probes during the fabrication procedure is made easy by remote interrogation and monitoring of optical fiber platforms. The optimized procedure is independent from the adhesion substrate onto the fiber tip. In fact, we have also successfully applied the optimized procedure for depositing the MGs layer onto the bulk optical fiber, i.e., directly onto the glass facet, where no gold layer was deposited. It's worth mentioning that the overall fabrication time needed to make an optrode, including MGs synthesis and probe fabrication, clearly depends on the choice of substrate. In fact, procedures such as titanium/gold coating and FIB patterning or silanization of bare fiber have different durations as specified in the materials and method section. Overall, the results presented in this work set the stage for the large scale production of MGs assisted optical fiber probes in industrial applications.

2. Materials and Methods

2.1. Microgel Synthesis and Characterization

For MG synthesis we used poly(Nisopropylacrylamide) (pNIPAm) that has emerged as the most popular polymer in the class of stimuli-responsive polymers [15] for their use in in sensing and biosensing applications [16–18]. The set of MGs has been prepared by following the same procedure described in [11]. Briefly, a water solution (98 mL) of N-isopropylacrylamide (NIPAM, 0.900 g), N,N'-methylenebis-(acrylamide) (BIS, 0.050 g), and sodium dodecyl sulfate (SDS solution 20% w/v,

25 µL) were heated to 70 °C under nitrogen atmosphere for 1 h. A potassium persulfate (KPS) initiator solution (0.050 g in 1 mL of milli Q water) was added to start the polymerization. After 15 min, acrylic acid solution (0.048 g in 1 mL of milli Q water) was then injected into the solution and polymerization proceeded for 5 h. The PNIPAm-co-AAc MGs were purified by using a dialysis tubes (12–14 k nominal MWCO) and finally lyophilized. The hydrodynamic radius of PNIPAm-co-AAc MGs in buffer solution at room temperature was 213 ± 4 nm.

On the other hand, the set of MGs deposited directly onto the tip of the bulk fiber has been synthesized in a different way in order to ensure their covalent binding on the glass surface (see Section 2.3 for further details). Specifically, NIPAM (0.475 g), BIS (0.053 g), and maleic acid (MAAC, 0.050 g) were dissolved in 90 mL filtered, deionized water in a three neck flask [19]. The solution was heated to 70 °C and purged with nitrogen. The polymerization was initiated by addition of ammonium persulfate (APS, 0.06 g) dissolved in 10 mL of water. After 6 h, the polymerization reaction was stopped and the colloidal suspension was cooled down rapidly below room temperature in an ice bath. The synthesized PNIPAm-co-MAAc MGs were purified by using a dialysis tubes (12–14 k nominal MWCO) so that the unreacted monomers, dissolved polymer chains, and unreacted initiator can be removed. Finally, the collected MGs were lyophilized. The hydrodynamic radius of PNIPAm-co-MAAc MGs in buffer solution at room temperature was 278 ± 5 nm.

2.2. LOF Probe Fabrication

First, a 2 nm thick titanium layer was deposited via electron beam evaporation (Sistec KL400C, Kenosistec Srl, Binasco, Italy) on the cleaved end of a standard single mode optical fiber (Corning SMF-28 9/125 Corning Incorporated, New York, NY, USA). Then, we deposited a gold layer with a thickness of 50 nm. The gold layer was patterned with a square lattice of holes in order to excite a localized plasmon resonance (LSPR). To set the resonance wavelength in the single mode operation regime of optical fibers, the lattice period and the hole radius were chosen to be 850 nm and 150 nm, respectively. The hole pattern was written by means of a focused ion beam milling process, in which we used the FEI Quanta 200 3D Instrument (Thermo Fisher Scientific, Waltham, MA, USA) with beam currents of 30 pA and accelerating voltages of 30 kV energy [8,20]. The time necessary to deposit Ti and Au is about 10 min and the FIB patterning takes less than 1 min.

2.3. Bare Optical Fiber Probe Silanization

The surface silanization is a crucial step to ensure the efficiency and reproducibility of the MGs immobilization step on the bare optical fiber probe. The hydroxyl-groups of the cleaved and cleaned fibers were activated by exposing them with piranha solution [Sulphuric acid (95%)/Hydrogen peroxyde 4/1 (v/v)] for 30 min at room temperature. The optical fiber was rinsed with water and left in contact with 10% (v/v) ethanolic solution of aminopropyl triethoxysilane (APTES) for 2 h at room temperature. The bare optical fiber probe was washed with water and ethanol to remove non-covalently adsorbed silane compounds and finally dried in a convection oven for 10 min at 120 °C. The above reaction produced free amine groups on the fiber surface able to couple to the carboxylic group on the MGs surface.

After the deposition of PNIPAm-co-MAAc MGs film (details on the deposition process are given in the next section), the immobilization of MGs on the NH$_2$ functionalized optical fiber was achieved by immersing the probe in MES buffer solution of N'-ethylcarbodiimide hydrochloride (EDC, 500 mM) for 10 h under magnetic stirring. Finally, the optical fiber was rapidly washed with water and then dried in the air for 10 min.

2.4. The Dip Coating Technique

The MGs coating procedure for the LOF probes is schematically shown in Figure 1: it consists of 4 main step: (i) dipping: by means of a dip coater (KSV NIMA KN4001, Biolin Scientific Oy, Espoo, Finland) the optical fiber probes are dipped into a 1.5 mL centrifuge tube containing a 500 µL aliquot of MGs solution at 5% with pH = 3. During the dipping phase, the temperature of MGs dispersion was kept at 10 °C by using a customized metallic holder heated by Peltier cells [8]. The speed and the duration of the dipping phase depend on the specific experiment carried out (details in the results section); (ii) first drying: after dipping, the fibers are taken out from the MGs solution and dried. Temperature and duration depend on the specific experiment carried out (details in the results section); (iii) rinsing: once dried, the optical fibers are immersed in a deionized water bath under magnetic stirring in order to ensure breaking up of potential multilayers and their removal from the tip surface. Temperature and duration are varied in our experiments carried out (details in the results section); (iv) second drying: after rinsing, the probes are dried again. Temperature and duration depend on the specific experiment (details in the results section).

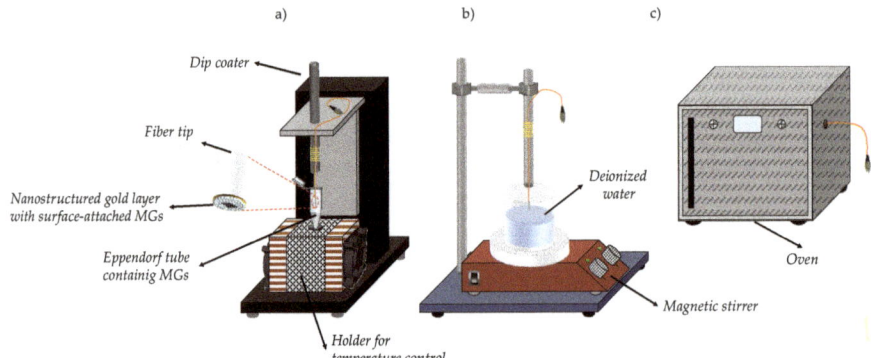

Figure 1. Schematic view of the dip coating procedure. (**a**) Dipping, (**b**) rinsing and (**c**) drying steps. The inset in (**a**) shows a schematic of the LOF probe integrated with MGs.

2.5. Optical Characterization

The reflection spectra of the fabricated probes were measured by using an optical setup described in [8]. Briefly, the fiber probes are illuminated by a broadband optical source (NKT SuperK COMPACT, NKT Photonics, Birkerød, Denmark) and an optical spectrum analyzer (Ando AQ 6715C, Yokogawa Electric Corporation, Tokyo, Japan) measures the reflected light thanks to a 2 × 2 directional coupler. Another spectrum analyzer measures the source light transmitted through the coupler. The reflectance spectrum is obtained by normalizing the sample spectrum with that of the source (taking also into account the transfer function of the coupler). Further details can be found in ref. [8].

2.6. Morphological Analysis

Morphological characterizations of the MGs film are directly performed onto the optical fibers tip by using the Atomic Force Microscope (AFM) (Agilent Technologies 5420, Agilent Technologies, Santa Clara, CA, USA) Specifically, AFM images are obtained by scanning the optical fiber tip in tapping mode in order to avoid damage of the polymeric film. The scanning of the fiber tip is achieved by using a customized holder that keeps the fiber in the right position [8]. The holder is essentially composed by an aluminum block where some grooves for the fibers are made by mechanical milling. A magnetic clamp lock the fibers vertically under the AFM scanner tip in such grooves. All the measurements were carried out with the MGs in the dry state. The raw data collected by the AFM are successively processed with the Pico Image software (Keysight Technologies, Santa Rosa, CA, USA).

3. Results and Discussion

3.1. Optical Monitoring of the Fabrication Process

We started our analysis by monitoring the reflection spectra of the LOF probes in all the deposition steps used in our previous procedure [11]. Such a procedure, customized to work on the optical fiber tip, was similar to that typically used for depositing MGs on planar substrates [21,22]. The details of the fabrication path, including time and temperature of all the steps previously described in Section 2.4, are summarized in Table 1.

Table 1. Fabrication step details of the previous deposition procedure.

Deposition Step	Time	Temperature
Dipping	1 h [1]	10 °C
First drying	1 h	45 °C
Rinsing	12 h	room
Second Drying	2 h	30 °C

[1] dipping speed 5 mm/min.

Figure 2a shows the reflection spectra of the LOF probe measured at the end of the different fabrication steps previously described in Section 2.3. Before the MGs deposition, the reflection spectrum of the LOF probe in air (dashed black curve) presents a dip centered at 1363 nm with a FWHM of 85 nm. The reflection dip is due to the excitation of a LSPR [8–12].

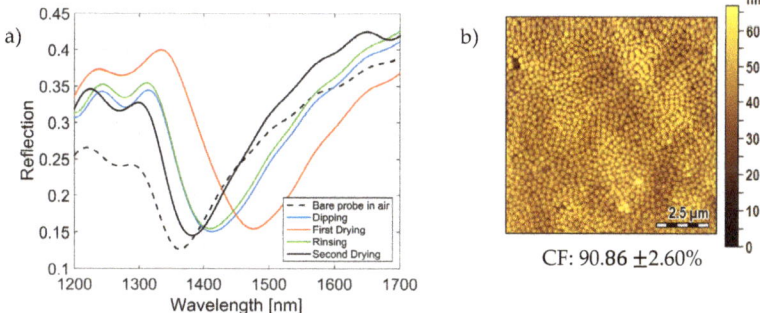

Figure 2. Characterization of the MGs film deposited following the previous procedure onto the goad coated optical fiber tip. (**a**) Reflection spectra of the LOF probes measured at the end of each fabrication step. (**b**) AFM characterization of the realized MGs film at the end of the entire procedure, pertaining to an unpatterned area of 10 × 10 µm close to the nanostructure.

When the probe is immersed in the MGs solution, the resonance (and thus the reflection dip) undergoes a red shift of ~54 nm (solid blue curve), coherently with an increase of the plasmonic mode equivalent refractive index [23]. Once the probe is extracted from the MGs solution, the resonant dip undergoes a further red shift of ~58 nm (solid orange curve), i.e., ~112 nm with respect to the initial spectrum pertaining to bare LOF probe. The strong red shift observed in response to a bulk refractive index decrease (from buffer solution to air) is principally due to the presence of a dried MGs multilayer (whose refractive index is about 1.47) attached onto the gold nanostructure. During the rinsing step, the resonant dip shifts back to lower wavelengths (coherently with the drying of the surface attached MGs), stabilizing at 1411 nm (green curve), i.e., ~6 nm lower with respect to the spectrum pertaining to the previous dipping step. This difference could be explained with the detachment of MGs which were not strongly adsorbed onto the gold substrate. Finally, after the second drying step, the resonant dip is subjected to a further blue-shift of ~30 nm (solid black curve). These data confirm that the MGs detached from the fiber tip surface during the rinsing step. The opposite direction of the wavelength shift with respect to the first drying step can be explained by considering that, in this case, the attached MGs give rise to a monolayer whose thickness is not thick enough to 'mask' the effect of the bulk medium. Overall, at the end of the entire process, the reflection spectrum is shifted by ~17 nm with respect to the bare LOF probe one.

At this stage, we carried out a morphological characterization of the MGs film realized onto the fiber tip. Without loss of generality, we scanned an unpatterned area of 10×10 µm close to the nanostructure realized in correspondence of the fiber core. The morphological analysis has been carried out only outside of the fiber core area, because the presence of the nanostructure makes difficult to evaluate the CF. However, as already demonstrated in our previous studies [8,11,12] the pattern does not alter the distribution of the MGs, which are still conformally deposited onto the gold nanostructure. The AFM image shown in Figure 2b confirms the creation of a uniform and compact MGs film (with an average height lower than 40 nm, compatible with the formation of a single layer [11]), characterized by a CF higher than 90%. The CF is defined as the ratio between the portion of substrate occupied by MGs and the whole substrate area; it is calculated by processing the AFM images, according to the same procedure described in our previous work [11]. Also, repeatability tests on the CF calculation have been carried out by using other 4 fibers, so that CF and the correspondent uncertainty have been evaluated as average and standard deviation among 5 probes. We remark that repeatability studies, over 5 samples, for the correct CF evaluation have been carried out for all the tests described in the rest of this work.

Figure 3 shows the evolution of the resonance wavelength (i.e., the wavelength corresponding to the reflection minimum) as a function of time during the dipping (Figure 3a), first drying (Figure 3b), rinsing (Figure 3c) and second drying (Figure 3d) step, respectively. Spectra acquisition interval was set to 1 min. For better reading the data, only the wavelength shifts pertaining to the first 20 min of each step are shown.

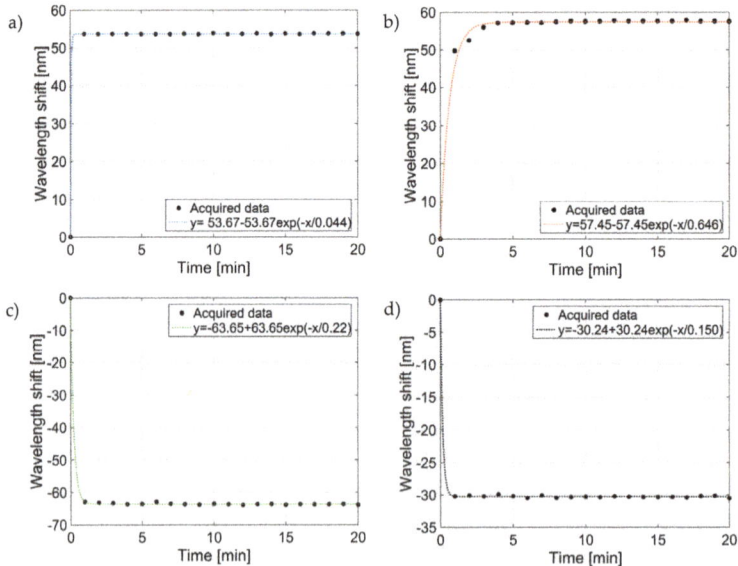

Figure 3. Resonant wavelength shifts measured during the fabrication steps of the previous deposition procedure. (**a**) Dipping step, (**b**) first drying step, (**c**) rinsing step, (**d**) second drying step. The dotted curves are the exponential fit of the acquired data.

Results essentially demonstrate that the dynamics associated to each fabrication step vanish in a time range significantly lower than that considered in the previous procedure. For estimating the extinction time of each dynamic, the wavelength shifts data were fitted with an exponential curve such as $y = a + b\,e^{-t/\tau}$. This allow us to find the time constants τ, and more specifically to define the interval (equal to $10 \times \tau$) required for the single fabrication step to reach the steady state condition. Consequently, the duration of the dipping phase can be reduced down to ~1 min; while the first drying step takes about 6.5 min to be completed. Furthermore, the rinsing step dynamics is going down after ~2 min, while a time interval of 1.5 min is enough for completing the second drying step. It is interesting to notice that the second drying dynamic takes a time interval that is about 4 times smaller than that of the first one, because of the smaller amount of MGs attached on the fiber probe surface.

The formation of the MGs layer in the solution cannot be correctly monitored just from the optical point of view (i.e., by monitoring only the reflection spectrum changes as a function of time). This is because when the MGs are swollen in the wet state, their refractive index basically corresponds to that of the liquid bulk solution in which they are immersed. For this reason, in the following sections, we also carry out a careful AFM morphological analysis for studying the influence that each fabrication step has on the characteristics of the MGs film created onto the fiber tip. The MGs layer achieved with the previous procedure shown in Figure 2b represents the benchmark for comparability purposes.

3.2. Study on the Dipping Step

We fabricated several probes by exploiting the previous procedure with the exception of the immersion speeds (details are summarized in Table 2). Specifically, 3 probes (i.e., the replica of the same LOF device) were dipped in the same MGs solution (pH 3, 10 °C) but with immersion speeds of 5, 100 and 200 mm/min. The speeds of 5 and 200 mm/min represent the minimum and the maximum value that it is possible to set with the dip coater. The immersion depth (2.5 mm) is the same for all the tests, so that the speeds of 5, 100 and 200 mm/min correspond to dipping time interval of 1, 0.05 and 0.025 min, respectively. The dipping time also includes the fiber extraction step. As mentioned before, both the rinsing and drying steps are carried out according to the previous procedure [11].

Table 2. Fabrication step details of the deposition procedure with the optimized dipping step.

Deposition Step	Time	Temperature
Dipping	1, 0.05, 0.025 min [1]	10 °C
First drying	1 h	45 °C
Rinsing	12 h	room
Second Drying	2 h	30 °C

[1] corresponding to dipping speeds of 5, 100 and 200 mm/min respectively.

Figure 4 shows the AFM images of the fiber tips of the fabricated samples. The relative CF evaluated at the end of the deposition procedure is reported in the same figures. The CFs are found to be larger than 90% in all the cases, with a standard deviation smaller than 5%. The data confirm that a dipping duration less than 1 min is enough for creating a compact and uniform MGs layer. However, in the cases of dipping speeds of 100 and 200 mm/min (Figure 4b,c), the quality of the MGs films appears slightly degraded due to the presence of some areas with a lower particle density, probably due to a too fast dipping speed. On the other hand, with the dipping procedure of 5 mm/min (Figure 4a), no substantial differences among the MGs film properties are found in terms of compactness and uniformity degree if compared with the benchmark in Figure 2b. In the wake of this consideration, for the following tests, we opted for a dipping time of 1 min.

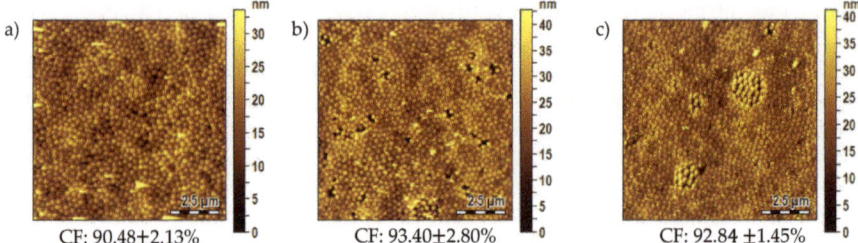

Figure 4. AFM images of the MGs film realized onto the gold coated fiber tip with different dipping time and speed. (**a**) 1 min (at 5 mm/min), (**b**) 0.05 min (at 100 mm/min), (**c**) 0.025 min (at 200 mm/min).

3.3. Study on the Drying Steps

After dipping, once the fiber probe is taken out from the MGs solution, it is necessary to dry the sample before the next rising step. As the MGs are sensitive to temperature, it is important to monitor the drying conditions (time and temperature). In our previous procedure, this step was particularly time consuming since the drying procedure was accomplished by means of an oven for 1 h at 45 °C. Analogously, after the rinsing step the MGs film is allowed to dry once again for 2 h at 30 °C in an oven before the optrode can be finally used for the specific application.

To understand to what extent it was necessary to control the drying conditions, we fabricated other probes carrying out the drying steps at room temperature (not controlled environment). In the wake of the results of Figure 4, discussed in the previous section, the dipping time was fixed at 1 min (with a speed of 5 mm/min). Coherently with the dynamic shown in Figure 3b,d, we investigated a drying duration of 10 min (for both the first and the second drying steps). Details on the deposition procedure are summarized in Table 3. The AFM image of the achieved MGs film (at the end of the whole deposition procedure) is shown in Figure 5.

Table 3. Fabrication step details of the deposition procedure with optimized drying steps.

Deposition Step	Time	Temperature
Dipping	1 min [1]	10 °C
First drying	10 min	room
Rinsing	12 h	room
Second Drying	10 min	room

[1] dipping speed 5 mm/min.

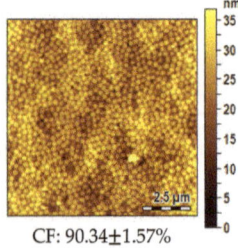

Figure 5. AFM image of the MGs film realized onto the gold coated fiber tip with a dipping time and drying steps reduced respectively to 1 min and 10 min, and rinsing time of 12 h.

From a visual comparison between the film in Figure 5 and the benchmark in Figure 2b, it is clear that 10 min are enough to ensure the correct drying of the MGs, and the consequent formation of a compact MGs monolayer. This aspect is confirmed by the value of the CF (also in this case evaluated as average among 5 probes fabricated with the same procedure) which is higher than 90%. This result confirms that it is possible to significantly reduce the duration of both the first and the second drying steps without using controlled temperature environment.

3.4. Study on the Rinsing Step

The rinsing step is necessary for removing the excess MGs particles which are not correctly deposited onto the fiber tip surface, thus causing the possible formation of a multilayer. To this end, the probes are immersed in water under magnetic stirring. Previous studies have demonstrated that the water temperature does not affect the MGs detachment from the substrate [24,25]. Following a similar approach exploited before (details in Table 4), we fabricated other probes with a rinsing time of 10 min, corresponding to the 1.4% of the duration considered in the previous procedure (12 h). Note that the new rinsing time is sufficiently larger with respect to the extinction time of 10τ extrapolated from the dynamic shown in Figure 3c.

Table 4. Fabrication step details of the optimized deposition procedure.

Deposition Step	Time	Temperature
Dipping	1 min [1]	10 °C
First drying	10 min	room
Rinsing	10 min	room
Second Drying	10 min	room

[1] dipping speed 5 mm/min.

The AFM image shown in Figure 6 demonstrates that 10 min are enough to break up and remove the MGs agglomerates, providing a compact and uniform monolayer with a CF of about 91%. From a one to one comparison between Tables 1 and 4 (and the relative AFM images shown in Figures 2b and 6) it results that the optimized procedure allows to achieve a saving of 98.3%, 83.3%, 98.6% and 91.6% of the time needed for the dipping, first drying, rinsing and second drying steps respectively. Moreover, the new procedure does not require the use of the oven (and related temperature and humidity controlled conditions) for the dying steps.

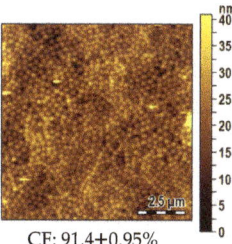

CF: 91.4±0.95%

Figure 6. AFM image of the MGs film realized onto the gold coated fiber tip with the optimized procedure (dipping: 1 min, drying: 10 min, rinsing: 10 min).

3.5. Remarks on the Optrode Performances

To understand if the time reduction affects the device performances, we carried out a thermal characterization on the optrode fabricated with the new optimized procedure whose details are resumed in Table 4. Details on the temperature measurements are reported in [11]. Figure 7 shows the evolution of the resonant wavelength (corresponding to the reflectance dip) as a function of temperature in the range 6–49 °C at pH 4.

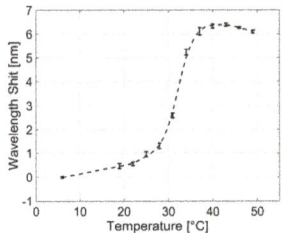

Figure 7. Resonant wavelength shift as function of solution temperature at pH 4 for the optrode fabricated with the new optimized procedure. The error bars represent one standard deviation evaluated on 5 acquisitions. The dashed lines were obtained with a smoothing spline fitting.

These results are quantitatively in line with those pertaining to the same optrodes fabricated with our previous procedure, which were discussed in [11].

3.6. Influence of the Substrate Typology

The results discussed so far referred to the integration of MGs layers onto the gold coated (patterned) optical fiber tip. To evaluate the generality of our finding, we studied also the case in which the MGs film is created directly onto the bare fiber tip, i.e., onto the silica glass. In this regard, it is important to underline that, in the case of a gold surface, the MGs-Au bond is favored by the Van der Waals interactions, and by the coordination between nitrogen and oxygen atoms [25–27]. Vice versa, the MGs-glass (SiO_2) interactions are weak and strongly depends from the solution pH and ionization state of silanol groups on glass surface [28]. For these reasons, to favor the MGs attachment onto glass surface, it is necessary to treat the bare optical fiber with silane and promote the covalent coupling of the MGs on the glass surface. To this end, after cutting, the fiber tip underwent a silanization process, and a new set of PNIPAm-co-MAAc MGs was synthesized as previously mentioned in Section 2.1. Details on the entire procedure are provided in Section 2.3.

For depositing the MGs film, we exploited the optimized procedure whose details are resumed in Table 4. The reflection spectrum of the silanized bare fiber is essentially flat (with a value of about 4%) simply because the silanization procedure does not affect the bare fiber behavior from the optical point of view. In other words, the glass surface modification is only 'chemical' and, obviously, no resonant effect are generated. Also in this case, the morphological analysis was carried out at the end of the second drying step. Figure 8 confirms the creation of a uniform and homogeneous layer of MGs with a CF of ~90%. Although not directly evaluable by means of optical characterization, the robustness of this optrode (in which the MGs are attached directly onto the uncoated fiber) is guaranteed by the silanization procedure, which allows to promote a covalent anchoring of the MGs on the glass surface. Moreover, there is no evidence to suggest that the MGs move or migrate from the fiber surface during rinsing under stirring, thus indicating a very strong MGs-substrate bond.

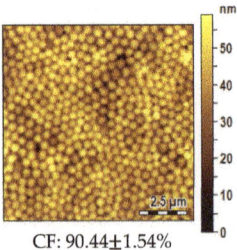

CF: 90.44±1.54%

Figure 8. AFM image of the MGs film realized onto the silanized glass fiber tip, following the optimized procedure.

4. Conclusions

In conclusion, we have proposed an optimized dip coating procedure for creating a uniform MGs layer onto the tip of optical fibers. This study complements our previous work in which we determined the optimum MGs dispersion parameters (solution temperature, pH and particle concentration) needed for achieving a compact MGs layer [11]. Here, we have essentially monitored the duration of the deposition steps and recognized that all of them can be shortened without affecting the quality of the MGs layer created on the fiber tip. Specifically, we have demonstrated that it possible to obtain MGs layers with coverage factors larger than 90% in about 30 min, which represents a significant step ahead with respect to the previous procedure, which was about 30 times longer. Our procedure is general and it has been successfully validated on both gold coated and uncoated optical fiber tips. Overall, our studies represent a solid foundation for developing advanced LOF probes integrated with MGs, with high fabrication throughput (the probe production can be also parallelized), and thus ready to be exploited at the industrial application level.

Author Contributions: L.S., M.G., A.M., collected and analyzed the data; A.A. and M.R. synthesized the MGs and silanized the optical fibers. L.S., M.G. and A.A. deposited the MGs films; E.B., V.L.F. and A.M. fabricated and characterized the optical fiber devices; A.R. conceived the idea and designed the experiments; A.C. supervised the project; all authors contributed to the writing of the manuscript.

Funding: This research received no external funding.

Acknowledgments: We thank Sofia Principe and Federica Gambino for their support during the MGs deposition onto the bare optical fiber.

Conflicts of Interest: The authors declare no conflict of interest.

References

1. Cusano, A.; Consales, M.; Crescitelli, A.; Ricciardi, A. *Lab-on-Fiber Technology*; Springer: New York, NY, USA, 2015; Volume 56.
2. Consales, M.; Ricciardi, A.; Crescitelli, A.; Esposito, E.; Cutolo, A.; Cusano, A. Lab-on-Fiber Technology: Toward Multifunctional Optical Nanoprobes. *ACS Nano* **2012**, *6*, 3163–3170. [CrossRef] [PubMed]
3. Ricciardi, A.; Consoles, M.; Quero, G.; Crescitelli, A.; Esposito, E.; Cusano, A. Versatile Optical Fiber Nanoprobes: From Plasmonic Biosensors to Polarization-Sensitive Devices. *ACS Photonics* **2014**, *1*, 69–78. [CrossRef]
4. Ricciardi, A.; Crescitelli, A.; Vaiano, P.; Quero, G.; Consales, M.; Pisco, M.; Esposito, E.; Cusano, A. Lab-on-fiber technology: A new vision for chemical and biological sensing. *Analyst* **2015**, *140*, 8068–8079. [CrossRef] [PubMed]
5. Vaiano, P.; Carotenuto, B.; Pisco, M.; Ricciardi, A.; Quero, G.; Consales, M.; Crescitelli, A.; Esposito, E.; Cusano, A. Lab on Fiber Technology for biological sensing applications. *Laser Photonics Rev.* **2016**, *10*, 922–961. [CrossRef]
6. Kostovski, G.; Stoddart, P.R.; Mitchell, A. The Optical Fiber Tip: An Inherently Light-Coupled Microscopic Platform for Micro- and Nanotechnologies. *Adv. Mater.* **2014**, *26*, 3798–3820. [CrossRef] [PubMed]
7. Carotenuto, B.; Micco, A.; Ricciardi, A.; Amorizzo, E.; Mercieri, M.; Cutolo, A.; Cusano, A. Optical Guidance Systems for Epidural Space Identification. *IEEE J. Sel. Top. Quantum Electron.* **2016**, *23*. [CrossRef]
8. Aliberti, A.; Ricciardi, A.; Giaquinto, M.; Micco, A.; Bobeico, E.; La Ferrara, V.; Ruvo, M.; Cutolo, A.; Cusano, A. Microgel assisted Lab-on-Fiber Optrode. *Sci. Rep.* **2017**. [CrossRef] [PubMed]
9. Giaquinto, M.; Micco, A.; Aliberti, A.; Ricciardi, A.; Ruvo, M.; Cutolo, A.; Cusano, A. Microgel Photonics and Lab on Fiber Technology for Advanced Label Free Fiber Optic Nanoprobes. *Proc. SPIE* **2016**. [CrossRef]
10. Ricciardi, A.; Aliberti, A.; Giaquinto, M.; Micco, A.; Cusano, A. Microgel Photonics: A Breathing Cavity onto OPTICAL FIBRE TIP. In Proceedings of the 24th International Conference on Optical Fibre Sensors, Curitiba, Brazil, 28 September–2 October 2015; SPIE: Bellingham, WA, USA, 2015.
11. Giaquinto, M.; Micco, A.; Aliberti, A.; Bobeico, E.; La Ferrara, V.; Menotti, R.; Ricciardi, A.; Cusano, A. Optimization Strategies for Responsivity Control of Microgel Assisted Lab-On-Fiber Optrodes. *Sensors* **2018**, *18*, 1119. [CrossRef] [PubMed]

12. Giaquinto, M.; Ricciardi, A.; Aliberti, A.; Micco, A.; Bobeico, E.; Ruvo, M.; Cusano, A. Light-microgel interaction in resonant nanostructures. *Sci. Rep.* **2018**, *8*, 9331. [CrossRef] [PubMed]
13. Pelton, R.; Hoare, T. Microgels and their synthesis: An introduction. In *Microgel Suspensions: Fundamentals and Applications*; Wiley: Hoboken, NJ, USA, 2011; pp. 1–32.
14. Pelton, R. Temperature-sensitive aqueous microgels. *Adv. Colloids Interface Sci.* **2000**, *85*, 1–33. [CrossRef]
15. Plamper, F.A.; Richtering, W. Functional Microgels and Microgel Systems. *Accounts Chem. Res.* **2017**, *50*, 131–140. [CrossRef] [PubMed]
16. Wei, M.L.; Gao, Y.F.; Li, X.; Serpe, M.J. Stimuli-responsive polymers and their applications. *Polym. Chem.* **2017**, *8*, 127–143. [CrossRef]
17. Islam, M.R.; Ahiabu, A.; Li, X.; Serpe, M.J. Poly-(N-isopropylacrylamide) Microgel-Based Optical Devices for Sensing and Biosensing. *Sensors* **2014**, *14*, 8984–8995. [CrossRef] [PubMed]
18. Islam, M.R.; Irvine, J.; Serpe, M.J. Photothermally induced optical property changes of poly-(N-isopropylacrylamide) microgel-based etalons. *ACS Appl. Mater. Interfaces* **2015**, *7*, 24370–24376. [CrossRef] [PubMed]
19. Dhanya, S.; Bahadur, D.; Kundu, G.C.; Srivastava, R. Maleic acid incorporated poly-(N-isopropylacrylamide) polymer nanogels for dual-responsive delivery of doxorubicin hydrochloride. *Eur. Polym. J.* **2013**, *49*, 22–32. [CrossRef]
20. Micco, A.; Ricciardi, A.; Pisco, M.; La Ferrara, V.; Cusano, A. Optical fiber tip templating using direct focused ion beam milling. *Sci. Rep.* **2015**, *5*, 15935. [CrossRef] [PubMed]
21. Schmidt, S.; Hellweg, T.; von Klitzing, R. Packing density control in P (NIPAM-co-AAc) microgel monolayers: Effect of surface charge, pH, and preparation technique. *Langmuir* **2008**, *24*, 12595–12602. [CrossRef] [PubMed]
22. Nerapusri, V.; Keddie, J.L.; Vincent, B.; Bushnak, I.A. Swelling and deswelling of adsorbed microgel monolayers triggered by changes in temperature, pH, and electrolyte concentration. *Langmuir* **2006**, *22*, 5036–5041. [CrossRef] [PubMed]
23. Spackova, B.; Wrobel, P.; Bockova, M.; Homola, J. Optical Biosensors Based on Plasmonic Nanostructures: A Review. *Proc. IEEE* **2016**, *104*, 2380–2408. [CrossRef]
24. Serpe, M.J.; Jones, C.D.; Lyon, L.A. Layer-by-layer deposition of thermoresponsive microgel thin films. *Langmuir* **2003**, *19*, 8759–8764. [CrossRef]
25. Sorrell, C.D.; Carter, M.C.; Serpe, M.J. A "paint-on" protocol for the facile assembly of uniform microgel coatings for color tunable etalon fabrication. *ACS Appl. Mater. Interfaces* **2011**, *3*, 1140–1147. [CrossRef] [PubMed]
26. Lu, Y.; Drechsler, M. Charge-induced self-assembly of 2-dimensional thermosensitive microgel particle patterns. *Langmuir* **2009**, *25*, 13100–13105. [CrossRef] [PubMed]
27. Iori, F.; Corni, S.; Di Felice, R. Unraveling the interaction between histidine side chain and the Au(111) surface: A DFT study. *J. Phys. Chem. C* **2008**, *112*, 13540–13545. [CrossRef]
28. Burmistrova, A.; Steitz, R.; von Klitzing, R. Temperature Response of PNIPAM Derivatives at Planar Surfaces: Comparison between Polyelectrolyte Multilayers and Adsorbed Microgels. *Chemphyschem* **2010**, *11*, 3571–3579. [CrossRef] [PubMed]

© 2018 by the authors. Licensee MDPI, Basel, Switzerland. This article is an open access article distributed under the terms and conditions of the Creative Commons Attribution (CC BY) license (http://creativecommons.org/licenses/by/4.0/).

Article

Fluoropolymer-Wrapped Conductive Threads for Textile Touch Sensors Operating via the Triboelectric Effect

Morgan Baima and Trisha L. Andrew *

Departments of Chemistry and Chemical Engineering, University of Massachusetts Amherst, Amherst, MA 01003, USA; morgan.baima@gmail.com
* Correspondence: tandrew@umass.edu; Tel.: +1-413-545-1651

Received: 3 May 2018; Accepted: 6 June 2018; Published: 11 June 2018

Abstract: Touch-sensitive electrical arrays are the primary user interface for modern consumer electronics. Most contemporary touch sensors, including known iterations of textile-based touch sensors, function by detecting capacitive changes within a circuit resulting from direct skin contact. However, this method of operation fails when the user's skin or the surface of the touch sensor is dirty, oily or wet, preventing practical use of textile-based touch sensors in real-world scenarios. Here, an electrically touch-responsive woven textile is described, which is composed of fluoropolymer-wrapped conductive threads. The fluoropolymer wrapping prevents contaminant buildup on the textile surface and also electrically insulates the conductive thread core. The woven textile touch sensor operates via surface potential changes created upon skin contact. This method of operation, called the triboelectric effect, has not been widely used to create textile touch sensors, to date. The influences of surface wetness and varying skin surface chemistry are studied, and the triboelectric textile touch sensors are found to be advantageously insensitive to these environmental variables, indicating that triboelectric textiles have promise for practical use as touch interfaces in furniture and interior design.

Keywords: textile electronics; plain weave; touch sensor; triboelectricity

1. Introduction

Touch-sensitive electronic devices serve as the primary user interface for a plethora of current technologies, such as portable computers, e-readers, smart phones, smart watches, fitness trackers, in-vehicle consoles, interactive display screens and touchpads [1]. Two main types of touch sensitive screens are used in these technologies: resistive touch sensors and capacitive touch sensors [1,2]. Resistive touch sensors are composed of two conductive sheets, arranged face-to-face and physically separated by a micron-length air gap. When a user touches this type of device, the two conductive sheets are physically brought into contact with each other due to the force associated with the touch interaction, and an electrical signal is recorded and processed by a relevant operating system. Capacitive touch screens are comprised of a single insulating substrate, such as glass or poly(ethylene terephthalate), coated with a patterned array of conductive electrodes on one side. A small voltage is applied across this electrode array, creating a uniform electric field across the opposite, uncoated surface of the insulating substrate. When a user touches this uncoated side, a capacitor is dynamically formed across the electrode–insulator–finger arrangement, since human skin is conductive. The dynamic change in device capacitance is then registered by a controller.

While selected flexible, touch-sensitive device architectures are known [3,4], translating these devices to fiber-based substrates is not straightforward. Selected touch-responsive textiles that produce a resistance change when an exposed conductive thread comes into direct contact with human skin have been previously reported [5–12]. However, this method of operation fails when the user's skin or the surface of the touch sensor is dirty, oily or wet, preventing practical use of textile-based touch sensors in real-world scenarios.

Recently there has been increased scientific interest in triboelectric generators (TEGs) due to their ability to convert small force inputs into an electrical (voltage and current) output. Because these devices operate by detecting the surface potential changes created upon contact and release of dissimilar surfaces (due to either the triboelectric effect or contact electrification), simple adjustments of device architecture can provide a wide range of functional technologies [13–15]. Here, we use the triboelectric effect (or, similarly, the contact electrification effect) to transduce touch events into electrical signals using a woven textile composed of fluoropolymer-wrapped [16] conductive fibers. The fluoropolymer wrapping prevents contaminant build up on the textile surface and also electrically insulates the conductive thread core. Triboelectric textile touch sensors are found to be advantageously insensitive to common environmental variables, indicating their promising use as touch interfaces in furniture and interior design.

2. Materials and Methods

Device Assembly: Threads used in this study were either stainless steel 2-ply conductive thread purchased from Adafruit (product 640) (New York City, NY, USA) with a reported resistivity of 51 Ohm/m and diameter of 200 micrometers, or 2-ply conductive silver-plated nylon thread purchased from LessEMF (Cat.#A304) (Latham, NY, USA) with a reported resistivity of 2 kOhm/m and a diameter of approximately 140 micrometers. The poly(tetrafluoroethylene) (PTFE) tape used in this study was low density (0.35 to 0.5 specific gravity) thread sealant tape with a width of 0.5 inches and a thickness of 88.9 nanometers. All woven devices were fabricated by folding the PTFE tape in half over the conductive thread or threads, and then twisting the tape with an electric drill. For clarity, these tape-covered threads are referred to as "strings". The total weight of each of the PTFE-wrapped strings was less than 5% greater than the starting mass of the naked conductive thread. Textile touch sensors were hand-woven on a small loom (Beadsmith brand) using a plain weave pattern, using one string as the warp and one string as the weft, or 20 × 60 strings in total.

Electrical Characterization: All measurements were performed using a Wavenow potentiostat/galvanostat by Pine Research or a Tektronix three-channel oscilloscope. Voltage outputs were obtained using the open circuit potential experiment and current outputs were obtained using the chronoamperometry experiment with the applied voltage set to zero volts. Conductive threads from either the warp or the weft were attached to the potentiostat using an alligator clip. "Dry" tests were performed by placing the devices in a sealed dessicator with Drierite desiccant overnight. Nitrogen gas was blown on the samples before storage in the dessicator and during the data collection upon immediate removal from the dessicator. Wet tests (deionized water, tap water, and NaCl salt solutions) were performed by spraying the devices with the respective wet solvent or solution and testing while the surface of the device was wet. For the NaCl salt solutions, 40 mmol/L solutions were made using deionized (D.I.) and tap water, respectively, as the solvent in order to mimic human sweat. For "wet hand" tests, the individual executing the test sprayed his or her hand with the wet solvent or solution, and then collected the data from the wet skin test onto a dry device. Samples were rinsed with deionized (D.I.) water and patted dry between all wet tests. Samples were also air-dried for five minutes after patting dry. Lotion tests were performed by applying lotion to both the user's hand and the sensor surface (separately). For the lotion-on-hand tests, a generous amount (0.32 g, allowed to absorb, and then re-applied for a total of 0.64 g) of Lubriderm Daily Moisture for normal to dry skin was applied to the hands and rubbed into the skin. To prevent extensive absorption, no wait time was allowed, and tests were carried out immediately after applying the lotion. For the lotion-on-sensor tests, a 0.32 g aliquot of lotion was spread on the sensor surface and repeated two times. The lotion was then wiped away from the sensor surface using a paper towel doused in either isopropanol or Clorox Green Works All-Purpose Cleaner.

3. Results and Discussion

3.1. Fluoropolymer-Wrapped Conductive Threads

Human skin is, on average, considered to possess a positive surface charge [1]. Therefore, to dynamically create a triboelectric generator upon skin contact, a complementary dielectric material carrying negative surface charge is needed. Poly(tetrafluoroethylene) (PTFE), a fluoropolymer, was chosen as the negative triboelectric material due to the strong electron affinity demonstrated by fluorinated polymers [17]. Furthermore, PTFE is a longstanding protective coating that prevents the acid/salt corrosion of, and biofilm formation on, various electronic components. In order to create a textile touch sensor, conductive threads were, first, tightly wrapped with PTFE tape (which is widely available because of its use as a waterproof, grease-free joint sealant in household plumbing) using an improvised yarning technique. Any commercially-available conductive thread can be used; in this report, two-ply stainless-steel yarns or silver-coated nylon yarns were used as the conductive core. Upwards of 40 feet of PTFE-covered thread are produced per yarning run. All exposed surfaces of the conductive thread were completely covered with PTFE tape, and no conductive surfaces were exposed along the length of the thread, with the exception of the ends that were left bare to allow electrical contact. The PTFE tape was in tight physical contact with the conductive thread core, allowing surface charge collection and transport to an external circuit. These PTFE-covered conductive threads were then plain woven using a tabletop loom to create a textile touch sensor.

3.2. Plain-Woven Triboelectric Textile Touch Sensors

Figure 1 illustrates the straightforward assembly of a plain-woven textile touch sensor created using PTFE-wrapped conductive threads. A picture of the completed 5 inch × 5 inch textile is provided in Figure 1b, and the operative mechanism for detecting skin contact is illustrated in Figure 1c. When the positively-charged surface of skin comes into contact with the negatively-charged PTFE surface of the textile, the equilibrium surface potential of the textile surface is perturbed due to surface charge transfer between the skin and the PTFE, and a compensating inductive current is generated through the conductive thread core. Upon separation of the skin from the PTFE, surface charge transfer ceases and a compensating current is created in the reverse direction to restore the equilibrium surface charge distribution of the textile.

The voltage generated by a touch action is determined by the amount of extra surface charges created upon skin/textile contact. Voltages between 4 and 5 volts are typically generated with a load resistance of 50 MΩ when a clean, dry finger contacts the dry textile touch sensor (or 1–2 V with a 500 GΩ load resistance). The current output is proportional to the amount of surface charge collected by the conductive thread core. A few microamperes of current are typically produced upon skin contact. The current output resulting from a single touch-release action on the woven textile touch sensor is depicted in Figure 1d.

Each warp thread in the woven textile touch sensor can, in theory, be connected to an independent sensing channel of a controller, which will only register an electrical signal when a user touches the part of the textile containing this thread. For this study, an oscilloscope with three independent channels was used and, therefore, the bare ends of the warp threads of the textile were portioned and bundled together to form three spatially-distinct (lengthwise) channels, labelled Region 1, 2, and 3 (Figure 2). A movie showing the electrical output from touching actions is provided in the Supporting Information (Video S1). Figure 2b shows groups of pulses generated in three different oscilloscope channels upon repeated touch–release actions performed on different spatial regions of the textile sensor. Each of the three channels operates independently and only manifests an electrical output when the corresponding region of the textile is touched. Therefore, this textile touch sensor automatically demonstrates spatial resolution concomitant with the weave density (or fill) used to create the device. A looser weave density would result in lower spatial resolution of touch events; therefore, only tight-woven textiles were investigated here. Further, when a finger is swiped across a large region of the textile, the shape of the electrical output is notably different and multiple channels record the dragging motion.

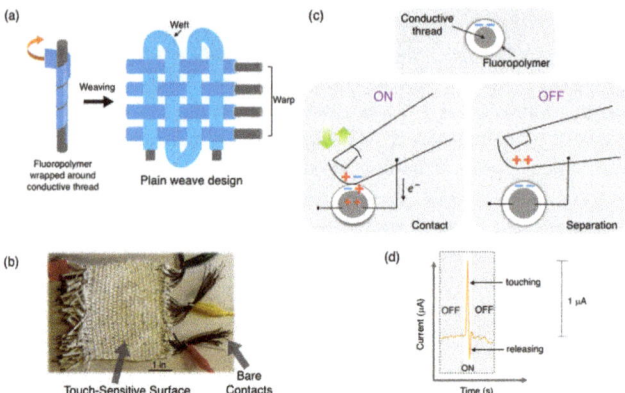

Figure 1. (a) Illustrated assembly of a woven textile touch sensor. The fluoropolymer is colored differently in the illustration for clarity. (b) Optical image of completed device. (c) Surface potential changes caused by skin contact and release. Not depicted to scale. (d) Current output of the textile touch sensor for a single touch and release action. The time duration of the "ON" signal is directly related to the duration of skin contact on the textile.

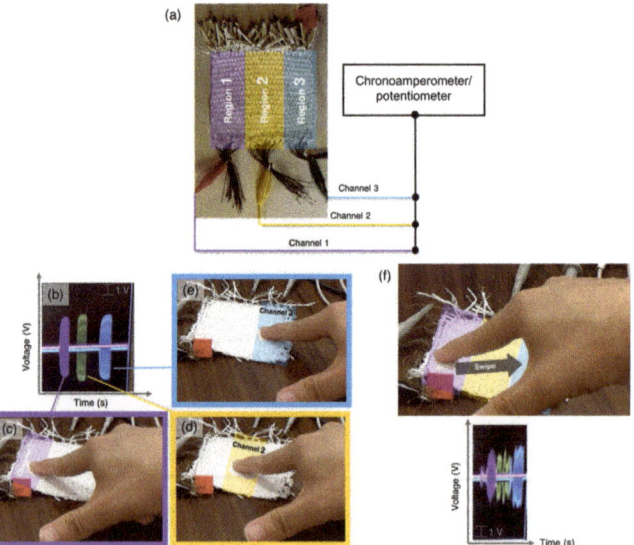

Figure 2. (a) Circuit diagram for testing woven textile touch sensors. The conductive threads from each string are bundled and connected to the potentiometer/chronoamperometer, segmenting the device into separate readout channels indicated by the different colored areas. (b) Voltage output with a 500 GΩ load resistance from three different sensing channels (purple, yellow, blue) when the corresponding spatial regions of the textile sensor (highlighted) are touched (c–e). (f) Voltage output with a 500 GΩ load resistance when a finger is swiped across the length of the textile touch sensor.

Current output is, generally, the limiting metric for all triboelectric devices, which otherwise generate remarkably high open-circuit voltages [13–16]. Therefore, in this study, the current output of the textile was primarily used as the distinguishing metric with which to identify touch events under various conditions.

Gestural differences result in varying current output from the textile touch sensor (Figure 3). Different touch interactions exert different characteristic forces on a surface [3,4]. A greater applied force results in a larger surface area of skin–textile contact, creating greater triboelectrification/contact electrification and resulting in an increased peak magnitude under short circuit conditions. The average current output created by a variety of common touch gestures on the woven textile sensor are plotted in Figure 3. Events with smaller applied force, such as swiping the hand across the device or touching softly, resulted in smaller short-circuit current outputs than harder slaps or pounds. Nevertheless, lighter touching actions still produced sizeable current output (few microamperes) that would be sufficiently detectable by existing electronic circuitry for interactive sensing applications.

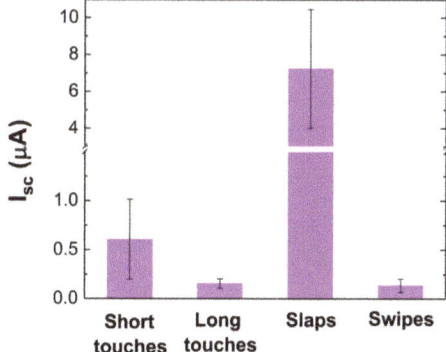

Figure 3. Current output under short circuit conditions for different gestures. Short touches and slaps lasted approximately 0.5 s, while long touches lasted approximately 1 to 1.5 s, and swipes lasted approximately 2 s.

Touch sensors are typically operated under dirty, "real-world" conditions where grease, sweat, salt, water and/or other biofilms accumulate on the surface of the sensor. Therefore, it is necessary to understand how the electrical output of textile touch sensors is affected by common environmental variables, such as water and sweat, that can be present on the surface of the interactor's skin. In order to assess the performance of the textile touch sensors in humid and sweaty environments, the current output under short circuit conditions was tested in two different ways: by wetting the sample surface and testing with dry hands and by using wet/salty hands to touch a dry sample surface. Both of these testing methods were explored because the PTFE wrapping effectively protects the conductive thread cores of the textile and prevents salt/biofilm accumulation on the textile surface.

Minimal changes in the average current output from the textile touch sensor were observed for either wet, dry or oily testing conditions (Figure 4). The observed voltage outputs from the textile touch sensor were similarly insensitive to the particular wetness or dirtiness of the surface or of the user's hand (Figure S1), although a globally-lowered voltage output was observed in the presence of excess water or ions, likely because skin contact-created surface charges were rapidly dissipated into the aqueous or ionic environment [17]. Similarly, wet surfaces and wet/salty hands resulted in a greater standard deviation in the magnitude of the observed current output (Figure 4b,c) but the average current output value was similar to the average current output when a pristine, dry textile was touched with clean, dry hands.

The two exceptions were the noticeably lower current output generated when the textile touch sensor was touched with a finger dipped into a salt water solution made with tap water or with lotion-saturated skin. Only the current output was affected in both cases, and the voltage outputs remained unchanged compared to other testing conditions. Further, fingers dipped in brine made with distilled water did not yield a lowered current output. We posit that these tap water and lotion outliers likely arise due to mechanical disruptions—the salty tap water and lotion make the surface of the user's skin especially sticky and subsequent touch interactions jostle the electrical connections attached to the textile touch sensor. This likely causes an increase in thread resistance that manifests as a lowered current output.

Overall, surface wetness and wet/sweaty fingers did not meaningfully change the current output generated by the textile touch sensor, qualifying this device for real-world operability.

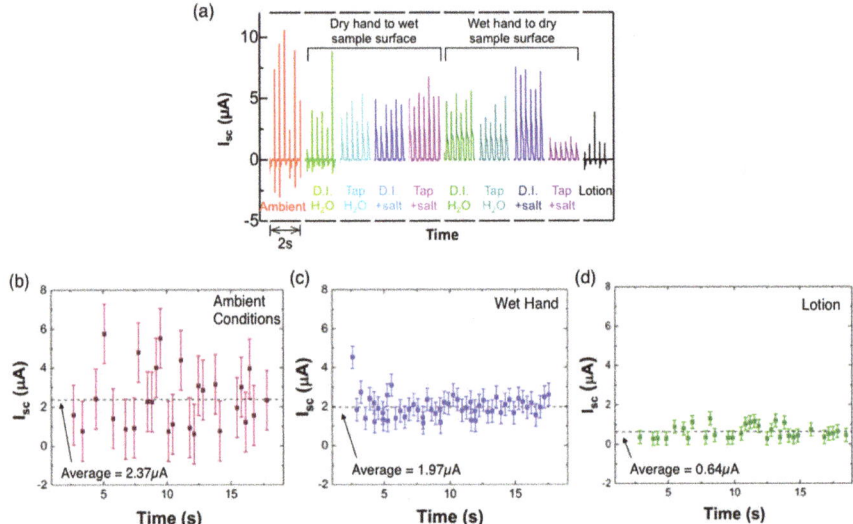

Figure 4. (a) Current output generated upon repeated skin contact and separation under different testing conditions. Peak values of current output with standard deviations for (b) Normal (dry sample, dry hands), (c) Wet hand touch actions, and (d) Touches with a small amount of lotion on skin. The dashed lines indicate the average peak value for each case.

To further explore the extent to which surface-accumulated oils affect the current output of the touch-responsive fabrics, a controlled sequence of experiments was performed before, during, and after lotion had been applied to the sensor's surface (not the user's hand) and subsequently wiped away using either a commercial cleaning product or isopropanol. Current outputs resulting from gentle touch actions (0.5 s duration) were recorded after each of the following actions were taken (corresponding to the experiment numbers depicted in Figure 5): (1) hands were washed with soap and water and a normal, dry sensor surface was touched; (2) excess lotion was applied to the skin surface and the sensor surface was touched; (3) lotion was coated on the sensor surface and then touched; (4) a second aliquot of lotion was applied to the sensor surface and touched; (5) hands were washed with soap and water and the sensor was touched; (6) the sensor surface was wiped with a paper towel doused with isopropanol and then touched; (7) the sensor surface was wiped again with another paper towel doused with isopropanol and then touched; and (8) the sensor surface was wiped a third time with a new paper towel doused with a commercial cleaning product.

The textile touch sensor showed an approximately ten-fold decrease in average current output when an excessive amount of lotion (representing extreme oil buildup) was present on the device surface but returned to normal functionality after the lotion was cleaned from the device surface using either a commercial cleaner or isopropanol. These results suggest that surface oils interfere slightly with the triboelectric effect, but not enough to prevent, or irreversibly alter, device operation. The maximum current output peak values and average current output peak values with standard deviations for these experiments are shown in Figure 5.

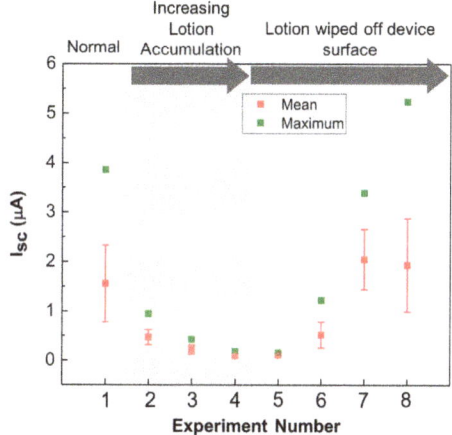

Figure 5. Maximum and mean values of current output generated upon repeated touch actions on a lotion-contaminated sensor surface. The experiment numbers are explained in the text.

4. Conclusions

The triboelectric effect was used to transduce touch events into an electrical signal using a plain-woven textile composed of fluoropolymer-wrapped conductive fibers. The textile touch sensor automatically demonstrated spatial resolution concomitant with the fill density. Gestural differences were also be identified by varying the current output from the textile touch sensor. The influences of surface wetness and varying skin surface composition were studied, and the textile touch sensors were found to be advantageously insensitive to these environmental variables. The woven textile touch sensors could foreseeably be implemented as a fabric touchpad to control smart garment functions or used as a control console in car seats and household furniture. Small patches of these touch-sensitive fabrics could also be sewn onto other backing textiles or incorporated into a garment, console, or furniture upholstery to serve as a fabric touchpad.

Supplementary Materials: The following are available online at http://www.mdpi.com/2079-6439/6/2/41/s1, Figure S1: Voltage output of the woven textile touch sensor under different, dirty operating conditions, Video S1: WovenTextileTouchSensor_LowRes.mp4.

Author Contributions: M.B. and T.L.A. conceived and designed the experiments; M.B. performed the experiments; M.B. and T.L.A. analyzed the data; T.L.A. and M.B. wrote the paper.

Acknowledgments: Financial support by the David and Lucille Packard Foundation is gratefully acknowledged.

Conflicts of Interest: The authors declare no conflict of interest.

References

1. Walker, G. A review of technologies for sensing contact location on the surface of a display. *J. Soc. Inf. Disp.* **2012**, *20*, 413–440. [CrossRef]
2. Baxter, L.K. *Capacitive Sensors: Design and Applications*; Wiley: Hoboken, NJ, USA, 1996.

3. Yuan, Z.; Zhou, T.; Yin, Y.; Cao, R.; Li, C.; Wang, Z.L. Transparent and flexible triboelectric sensing array for touch security applications. *ACS Nano* **2017**, *11*, 8364–8369. [CrossRef] [PubMed]
4. Pu, X.; Liu, M.; Chen, X.; Sun, J.; Du, C.; Zhang, Y.; Zhai, J.; Hu, W.; Wang, Z.L. Ultrastretchable, trasparent triboelectric nanogenerator as electronic skin for biomechanical energy harvesting and tactile sensing. *Sci. Adv.* **2017**, *3*, e1700015. [CrossRef] [PubMed]
5. Wijesiriwardana, R.; Mitcham, K.; Hurley, W.; Dias, T. Capacitive fiber-meshed transducers for touch and proximity-sensing applications. *IEEE Sens. J.* **2005**, *5*, 989–994. [CrossRef]
6. Vallett, R.; Young, R.; Knittel, C.; Kim, Y.; Dion, G. Development of a carbon fiber knitted capacitive touch sensor. *MRS Adv.* **2016**, *1*, 2641–2651. [CrossRef]
7. Roh, J.S. Textile touch sensors for wearable and ubiquitous interfaces. *Text. Res. J.* **2013**, *84*, 739–750. [CrossRef]
8. Ferri, J.; Lidon-Roger, J.V.; Morena, J.; Martinez, G.; Garcia-Breijo, E. A wearable textile 2D touchpad sensor based on screen-printing technology. *Materials* **2017**, *10*, 1450. [CrossRef] [PubMed]
9. Takamatsu, S.; Yamashita, T.; Imai, T.; Itoh, T. Fabric Touch Sensors using Projected Self-Capacitive Touch Technique. *Sens. Mater.* **2013**, *25*, 627–634.
10. Büscher, G.H.; Kõiva, R.; Schürmann, C.; Haschke, R.; Ritter, H.J. Flexible and Stretchable Fabric-Based Tactile Sensor. *Robot. Autonom. Syst.* **2015**, *63*, 244–252. [CrossRef]
11. Suen, M.; Lin, Y.; Chen, R. A Flexible Multifunctional Tactile Sensor using Interlocked ZnO Nanorod Arrays for Artificial Electronic Skin. *Proc. Eng.* **2016**, *168*, 1044–1047. [CrossRef]
12. Tu, H.; Chen, X.; Feng, X.; Xu, Y. A post-CMOS compatible smart yarn technology based on SOI wafers. *Sens. Actuators A* **2015**, *233*, 397–404. [CrossRef]
13. Wang, X.; Wang, Z.; Yang, Y. Hybridized nanogenerator for simultaneously scavenging mechanical and thermal energies by electromagnetic-triboelectric-thermoelectric effects. *Nano Energy* **2016**, *26*, 164–171. [CrossRef]
14. Lee, K.; Yoon, H.; Jiang, T.; Wen, X.; Seung, W.; Kim, S.; Wang, Z.L. Fully packages self-powered triboelectric pressure sensor using hemispere-array. *Adv. Energy Mater.* **2016**, *6*, 1502566. [CrossRef]
15. Wang, X.; Zhang, H.; Dong, L.; Han, X.; Du, W.; Zhai, J.; Pan, C.; Wang, Z.L. Self-powered high-resolution and pressure-sensitive triboelectric sensor matrix for real-time tactile mapping. *Adv. Mater.* **2016**, *28*, 2896–2903. [CrossRef] [PubMed]
16. Zhang, L.; Yu, Y.; Eyer, G.P.; Suo, G.; Kozik, L.A.; Fairbanks, M.; Wang, X.; Andrew, T.L. All-textile triboelectric generator compatible with traditional textile process. *Adv. Mater. Technol.* **2016**, *1*, 1600147. [CrossRef]
17. Thomas, S.W., III; Vella, S.J.; Dickey, M.D.; Kaufman, G.K.; Whitesides, G.M. Controlling the kinetics of contact electrification with patterned surfaces. *J. Am. Chem. Soc.* **2009**, *131*, 8746–8747. [CrossRef] [PubMed]

 © 2018 by the authors. Licensee MDPI, Basel, Switzerland. This article is an open access article distributed under the terms and conditions of the Creative Commons Attribution (CC BY) license (http://creativecommons.org/licenses/by/4.0/).

Review

Electrically Conductive Coatings for Fiber-Based E-Textiles

Kony Chatterjee, Jordan Tabor and Tushar K. Ghosh *

Department of Textile Engineering, Chemistry and Science, Wilson College of Textiles, North Carolina State University, Raleigh, NC 27606, USA; kchatte@ncsu.edu (K.C.); jatabor@ncsu.edu (J.T.)
* Correspondence: tghosh@ncsu.edu; Tel.: +1-919-515-6568

Received: 26 March 2019; Accepted: 22 May 2019; Published: 1 June 2019

Abstract: With the advent of wearable electronic devices in our daily lives, there is a need for soft, flexible, and conformable devices that can provide electronic capabilities without sacrificing comfort. Electronic textiles (e-textiles) combine electronic capabilities of devices such as sensors, actuators, energy harvesting and storage devices, and communication devices with the comfort and conformability of conventional textiles. An important method to fabricate such devices is by coating conventionally used fibers and yarns with electrically conductive materials to create flexible capacitors, resistors, transistors, batteries, and circuits. Textiles constitute an obvious choice for deployment of such flexible electronic components due to their inherent conformability, strength, and stability. Coating a layer of electrically conducting material onto the textile can impart electronic capabilities to the base material in a facile manner. Such a coating can be done at any of the hierarchical levels of the textile structure, i.e., at the fiber, yarn, or fabric level. This review focuses on various electrically conducting materials and methods used for coating e-textile devices, as well as the different configurations that can be obtained from such coatings, creating a smart textile-based system.

Keywords: flexible electronics; smart textiles; conductive coatings; e-textiles

1. Introduction

In recent years, electronic textiles (e-textiles) as a class of soft or flexible electronics have generated a growing interest due to their many potential applications in healthcare, security, entertainment, and others. E-textile systems are produced through the integration of various electrical devices, such as sensors [1–3], transistors [4–7], communication devices [8–10], energy harvesting and storage devices [11–14], and actuators [15–18], with textiles. As a hierarchical structure, textiles offer unique opportunities to integrate electrical functionalities at various levels—from fibers, yarns, to the finished product [19] (see Figure 1). While electronic capabilities can be integrated into any of these hierarchical levels, the integration of electrically conductive materials at the fiber level arguably enables the most seamlessly-integrated e-textile products. Fiber level integration of electrical capabilities is more likely to help retain the intrinsic textile characteristics of strength, flexibility, durability, comfort, etc., enhance the functionality by enabling communication, as well as sense and respond to the external environment [20].

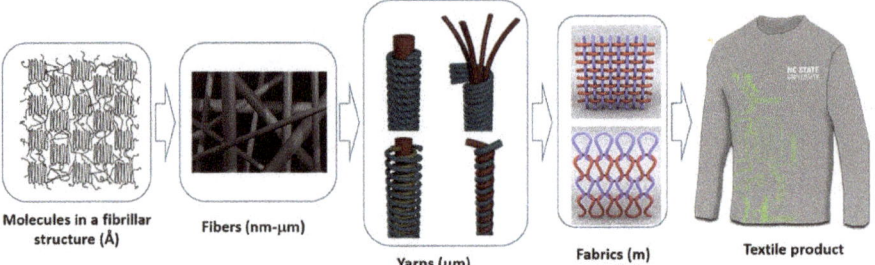

Figure 1. The hierarchy of textile structures as they progress from long chain polymers to the final textile product. Reproduced with permission from [19]. Copyright 2018, John Wiley and Sons: Hoboken, NJ, USA.

To achieve this goal of unobtrusive integration of electronics with textiles, it is important to use materials and methods that can impart the necessary electrical conductivity to a textile fiber. Electrical conductivity at the fiber level can be achieved by using either intrinsically conducting polymers (ICPs) in forming the fiber or by coating conventional insulating fibers with conducting materials [21]. Various materials such as ICPs [22,23], conducting polymer composites [24,25], metals [21,26], and carbon based materials, such as carbon nanotubes [27,28], carbon nanopowders [1] and graphene [29,30], have been used to achieve this. These materials have been applied using coating methods such as electro- and electroless deposition [31–33], dip-coating [34,35], and chemical vapor deposition (CVD) [23,30] to achieve such electrically conductive e-textile coatings. The electrically conductive coated fibers can be used in a variety of applications in e-textiles. However, we limit the scope of this review to cover one of the most explored areas in e-textiles—the fiber-based sensors [25–27]. A sensor is a device that can respond to an external stimulus by generating a measurable signal. Fiber-based sensing devices have been proposed for many applications, such as biomedical monitoring [22,28], security [29], sports [30], and others [24,36,37]. Such sensors are flexible, lightweight, conformable, and can be woven or knitted into various textile structures for facile integration of sensors within the fabric structure [19]. In principle, fiber-based sensors can be designed to operate using many of the principles of sensing, such as capacitive, resistive, piezoelectric, etc., to detect various stimuli such as pressure, strain, proximity, and temperature [28].

In this review, electronic capabilities integrated into the fiber-based sensors via any of the coating processes are explored. The first section explores the materials that have been commonly used for imparting such electrical conductivity to e-textile fibers, including ICPs, conducting polymer composites, metals, and carbon nanotube-based coatings, as well as the methods that have been employed to apply these coatings conformably onto the fibers. The following section covers types of coated fiber-based sensors classified based on their sensing principle as well as their performance and applications. While fibers and yarns differ considerably in textile processing, in this review, both fiber and yarn-based sensors are included.

2. Materials and Methods Used for Coating

Coating fibers or yarns with electrically conductive materials offers the inherent advantage of providing a route to transform already manufactured textile materials into electrically conductive e-textiles. This eliminates the need to add electrical functionality during fiber manufacturing, which many electrically conductive materials may not be compatible with [38]. Moreover, due to the various coating methods available, a variety of materials can be incorporated into the fiber without the limitation of using conventional textile processing techniques [39]. Coating processes also enable the incorporation of materials into e-textiles that cannot themselves form fiber, yarn, or fabric structures, such as ICPs, which are mechanically too weak to withstand various textile processes [39]. The subsequent sections explore the various materials and methods that have been used to render e-textile fibers conductive, with emphasis on limiting the discussion to fibers and yarn structures rather than complete fabrics.

2.1. Intrinsically Conducting Polymers (ICPs)

Research into ICPs began in 1974, when Shirakawa et al. first reported electrically conductive films composed of polyacetylene made via the polymerization of acetylene with 1000 times higher amounts of catalyst than usual [40]. ICPs can be classified as either cation or anion salts of highly conjugated polymer structures. The cation salts can be formed via electrochemical polymerization or chemical oxidation, whereas the less stable anion salts can be formed via electrochemical reduction or by using reagents [41]. While ICPs are generally semiconductors in their pristine state, they can be made more conductive by doping using either p- or n-dopants. Dopant molecules are introduced to add or remove electrons from the backbone of ICPs, resulting in increased conductivity of these polymers. ICPs do, however, have high redox activity and electron affinity, unlike conventional polymers [42]. The charge transport properties of ICPs depend on the packing of their polymer chains and their degree of order, and on the amount of impurities and structural defects present within the structure. In the case of polyacetylene—the simplest linear conjugated ICP—the conductivity is due to the strong interchain interaction, resulting in a fibrillar crystal structure consisting of π-stacked polymer chains. Moreover, due to the π-conjugation, ICPs such as polyacetylene possess a planar structure attributed to the sp^2 hybridization of the carbon atoms in the polymer chain, which can be modified by the introduction of dopants [43]. Other than polyacetylene, some of the most important classes of ICPs are polyparaphenylene vinylenes, polyethylene dioxythiophenes, polypyrroles (PPy), and polyaniline (PANI). Figure 2 shows the various ICPs as well as their band gap values (energy difference between the top of the valence band and the bottom of the conduction band).

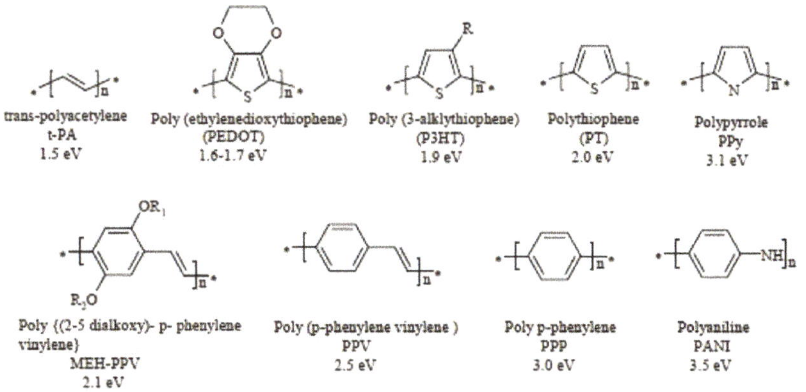

Figure 2. Chemical structures of various intrinsically conducting polymers (ICPs) with their corresponding band gap values. Reprinted with permission from [44]. Copyright 2009, Emerald Publishing Limited: Bingley, West Yorkshire, UK.

Since ICPs can be synthesized precisely in a controlled manner with tunable conductivity due to doping, and they have a variety of electrically, structural, and optical properties, they are promising materials for use in various flexible electronics applications [43]. However, due to their rigid backbone structure, which inhibits their solubility as well as their tendency to decompose at temperatures lower than their melting points, most ICPs cannot be melt processed via conventional textile methods [43]. Hence, using ICPs as coatings for e-textile applications is a viable technique to combine the unique properties of these materials with the flexibility, the strength, and the drapability of textiles [45]. Dip-coating is a facile method of applying ICPs onto textile fibers without damaging them in the process. Without the need for specialized equipment, it is an accessible process for making electrically conductive yarns—as long as the ICPs are obtainable in a solution form [46]. Kim et al. explored the formation of electrically conductive polyethylene terephthalate (PET) yarns using dip-coating to apply PANI onto these yarns [47]. PANI was converted into a conductive solution by functionalizing with dodecylbenzene sulfonic acid (DBSA). The proton in the acid reacts with the imine in the PANI, resulting in its protonation and rendering it conductive. The organic group within PANI is compatible with xylene, thereby forming a solution of DBSA-doped solution in xylene. PET yarns were then dip-coated in this solution in a coagulation bath to allow an electrically conductive coating to form on the yarns. This is possible due to the tendency of PANI to spontaneously agglomerate into film-like structures when exposed to a solid/solution interface during the coating process. The resistances of these yarns varied from 10^3–10^6 Ω, much higher than those obtained for conductive fibers made by melt spinning of polypropylene (PP) with PPy, which was attributed to the problems of structural inhomogeneity due to the melt spinning process. Moreover, coated yarns were able to better retain their original strength and flexibility.

Mostafalu et al. also used dip-coating to apply conductive PANI ink (along with other inks that were carbon nanopowder based or a composite of carbon nanopowders/PANI) onto cotton threads by sequentially passing these through multiple coating baths, with the ability to process many meters of thread via this process [48], as shown in Figure 3. Dryers were used after coating cycles to adhere the coating to the yarns, and these were subsequently collected on rotating spools, as shown in Figure 3a. PANI was able to form a three-dimensional (3D) network of nanofibers on the cotton yarns (Figure 3d). This enabled enhanced mechanical flexibility of the cotton yarn and more robust coating. These PANI coated cotton yarns were then used as pH sensor electrodes due to the biocompatibility offered by PANI and its high electrical conductivity. However, it is important to note that, while dip-coating is a relatively easy method, it does have certain limitations in terms of the uniformity of the conductive coating thickness, as well as the surface roughness and the agglomeration of conductive particles on the yarns. The latter can cause significant problems for electrical conduction and may even result in electrical discharge at these charge concentration zones [46]. Hence, a more conformal, controllable, and uniform method of applying ICPs as coatings is necessary for long-term, sustainable performance of e-textiles.

To overcome the challenges of dip-coating, alternative approaches such as in-situ polymerization have been explored to deposit ICPs onto textiles [45]. This process involves the physical adsorption of the ICP onto the surface of the textile, followed by polymerization along the plane of the solid–liquid interface to produce a thin, uniform coating that adheres to the fiber surface [49]. Yue et al. explored this technique by coating nylon-lycra fibers by depositing PPy onto the textile fabric substrate with an oxidant and a dopant to enable polymerization of the monomer. This formed conformal PPy coatings on all the fibers with a surface resistance of 149 Ω/square [50]. Interestingly, Yue et al. observed an initial slight increase and a subsequent decrease in resistance with increasing strain. This was attributed to the slow strain recovery of the nylon-lycra fabric at large elongations (50–100%), as well as the type of PPy formed using the specific dopants and the reaction conditions [50]. While this is an example of a fabric-based coating, it still involves the coating of individual fibers via in-situ polymerization. Hence, depending on the construction of the textile as well as the conditions used for polymerization, the electrical performance of ICPs can be tailored to fit the requirements of the sensing system. Sarvi et al. utilized this method to create conductive PANI coatings on multi-walled carbon nanotubes (MWNTs), forming MWNT-PANI core-shell nanofibers that were subsequently used as conductive nanofillers to enable conduction between poly(vinylidene fluoride) (PVDF) nanofibers [51]. PANI enables a better dispersion of MWNTs and enables the formation of conductive bridges between the PVDF fibers. Moreover, since the PVDF nanofibers are subsequently electrospun to form a conductive mat, it is important to form conductive links using the PANI-MWNT network, since air gaps between uncoated PVDF nanofibers prevent the formation of a conductive mat.

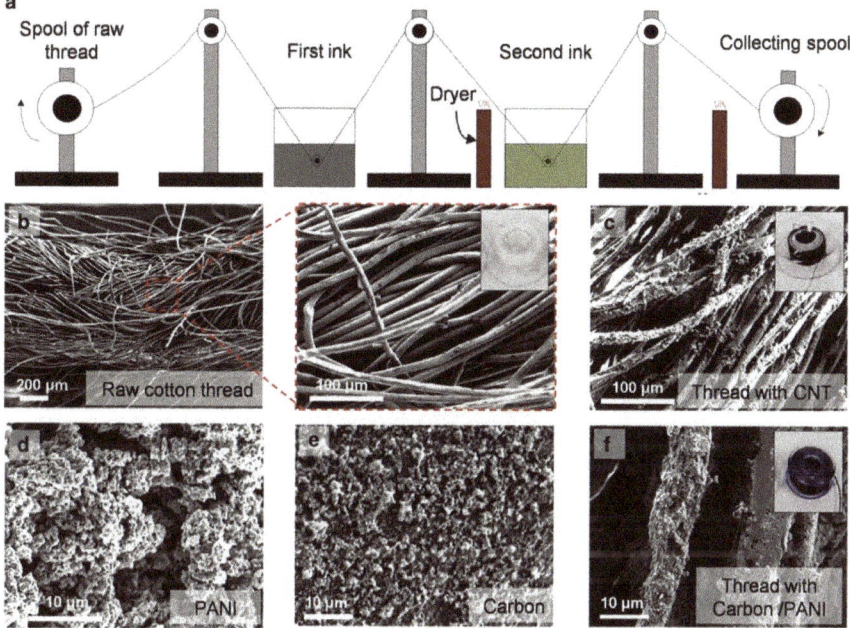

Figure 3. Coating cotton threads with electrically conductive materials to produce e-textile fibers for subsequent use as strain and temperature sensors: (**a**) A schematic of the coating setup used to dip coat and collect threads with the bath containing coating ink; (**b**) SEM image of the raw cotton thread before coating; (**c**) SEM image of a carbon nanotube (CNT) infused cotton thread after coating; (**d–f**) SEM images of polyaniline (PANI)-, carbon-, and carbon-PANI coated cotton threads, with the formation of a three-dimensional (3D) network of nanofibers of PANI shown in (**d**). Reprinted with permission from [48]. 2016 CC-BY 4.0 license, Nature Publishing Group: London, UK.

Eom et al. developed fiber based strain sensors using poly(3,4-ethylenedioxythiophene) (PEDOT) coated polyester (PS) fiber by directly polymerizing the (3,4-ethylenedioxythiophene) (EDOT) monomer to form the PEDOT coating [52], as shown in Figure 4. By making a monomer and an oxidant solution and then dipping the PS fiber into it for 20 min at 70 °C, they were able to produce a coated fiber with low resistance. Moreover, a poly(methyl methacrylate) (PMMA) coating was also added to the fiber to enable mechanical robustness and prevent the coating from damaging. In this manner, a PEDOT thickness of 100–300 nm was achieved with an electrical resistance of 600 Ω/cm. While the Young's modulus of the coated fiber was slightly higher, the overall mechanical behavior was comparable to uncoated PS fibers (Figure 4c). These yarns were subsequently used as strain sensors as knitted fabrics with interlocking insulating PS fibers to hold them in place (Figure 4e,f). The performance of these sensors is further discussed in Section 3.3.1. However, this technique does have certain limitations in terms of the inability to control the mass of coating formed during the in-situ polymerization reaction, problems such as delamination occurring after coating [52], as well as the speed with which the process proceeds. Moreover, due to its use of acidic oxidants and radical cations, this technique can also be harsh on the fabric substrates themselves—in some cases by dissolving the underlying yarn of fiber structure and resulting in film-like substrates [53].

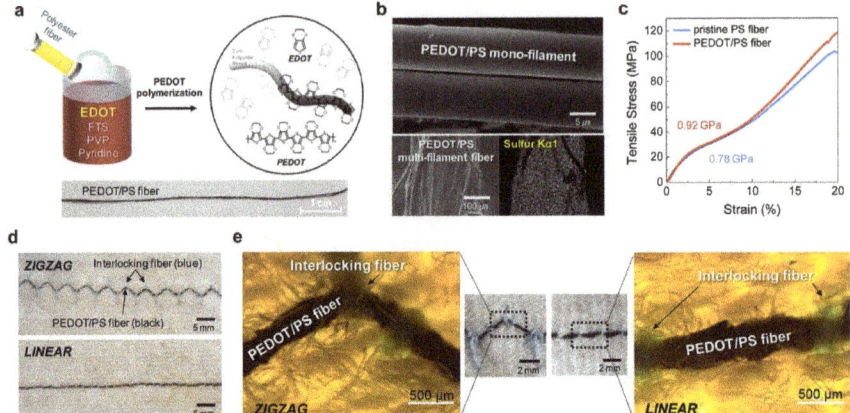

Figure 4. (**a**) Schematic illustrating the deposition of poly(3, 4-ethylenedioxythiophene) (PEDOT) on polyester (PS) fiber via (3,4-ethylenedioxythiophene) (EDOT) polymerization: Ferric *p*-toluene sulfonate (FTS) is used as an oxidant, whereas poly(vinyl pyrrolidone) (PVP) is used as a binder, and pyridine is used as a polymerization retarder; (**b**) SEM images and X-ray spectroscopy mapping of sulfur (S) Kα1 of PEDOT coated PS fiber; (**c**) stress–strain behavior for pristine polyester compared to PEDOT coated PS shows a good correlation, with the coated fiber having a larger Young's modulus due to the higher stiffness of the PEDOT coating; (**d**) blue colored coated fiber integrated into polyester fabrics in either zigzag or linear manners; (**e**) and (**f**) interlocking insulating polyester fibers used to keep the strain sensors in place. Reprinted with permission from [52]. Copyright 2017, American Chemical Society: Washington, DC, USA.

Vapor phase deposition (VPD) techniques such as CVD can combine the synthesis of the ICP and the deposition onto the textile substrate into a single step process without affecting the intrinsic drapability and the mechanical strength of the textile substrate [54,55]. In the CVD technique, a substrate is treated with an oxidant and then exposed to a monomer vapor, which subsequently polymerizes into a coating. This is followed by rinsing with a solvent such as methanol to remove unreacted monomers and any residual oxidants or byproducts [56]. Vapor deposition techniques can be divided into two types depending on the polymerization technique used to form the subsequent ICP film: (i) chain-growth polymerization methods, which include CVD, plasma-enhanced CVD (PECVD), initiated CVD (iCVD), and photoinitiated CVD (piCVD), and (ii) step-growth polymerization methods, which include oxidative CVD (oCVD) and the conventional vapor phase deposition (VPP) method [55]. VPD techniques enable the formation of a conformal coating by supplying monomers to the substrate's surface in the vapor phase, with the monomer self-initiating or introducing a second initiating species into the polymerization process to form a polymer film. The rate of this technique can be controlled by decreasing the substrate temperature in cases of ICPs and other polymeric films, thereby making this technique suitable for textile substrates that are temperature sensitive [55].

Bashir et al. studied the deposition of PEDOT on viscose and polyester yarns to form conductive fibers using oCVD to deposit the PEDOT [23,56], as shown in Figure 5. By increasing the amount of oxidant (FeCl$_3$) concentration, they were able to obtain conductive, thick PEDOT coatings on the viscose fibers (Figure 5e–i), but this also decreased the tenacity of the viscose fibers. Hence, to overcome this challenge of reduced mechanical properties, PET was explored as an alternative substrate for oCVD of PEDOT, especially due to its mechanical properties and its chemical stability to acidic environments (as presented by the FeCl$_3$ oxidant) compared to viscose fibers. PEDOT coated PET fibers exhibited increased electrical conductivity coupled with better mechanical properties than PEDOT coated viscose fibers, thereby enabling the formation of highly conductive fibers that retain their mechanical properties [32]. VPD techniques offer various advantages for ICPs, such as substrate independent deposition, conformable coatings, low temperature processing, and using ICPs that are difficult to process in solution form. However, the ability to scale up the organic VPD is still being explored, since most commercial VPD processes are designed for inorganic materials, which require high temperatures, input powers, and are incompatible with fragile organic components [55]. Nevertheless, research is being carried out in the field of scalability of organic systems VPD with batch reactors that can coat larger (>1 m) substrate areas [54,55]. The challenge still remains in terms of ensuring uniform deposition of the monomers and the oxidants on such large substrates [56,57].

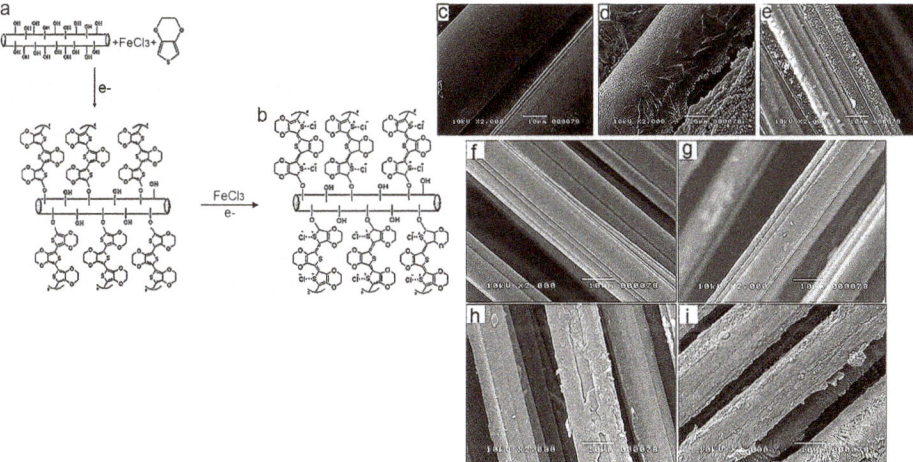

Figure 5. oCVD process to coat viscose and polyethylene terephthalate (PET) fibers with PEDOT: (**a**) Oxidation of the EDOT monomer to form PEDOT with the viscose fiber containing –OH functional groups that can bond with PEDOT; (**b**) Doping PEDOT with $FeCl_3$ to render PEDOT molecules electrically conductive by inducing positive charges on the polymer backbone; (**c**) Pristine viscose fiber; (**d**) PEDOT-coated viscose fiber; (**e**) PEDOT-coated PET fiber; for Figures (**c–e**), the scale bar is 10 µm. Figures (**f–i**) show SEM images of PEDOT coated viscose fibers with various concentrations of $FeCl_3$ at (**f**) 0 wt.%; (**g**) 3 wt.%, (**h**) 5 wt.%; and (**i**) 9 wt.%; scale bar for Figures (**f–i**) is 10 µm. Figures (**a–e**) reprinted with permission from [23]. Copyright 2011 John Wiley & Sons: Hoboken, NJ, USA. Figures (**f–i**) reprinted with permission from [56]. Copyright 2011 John Wiley & Sons: Hoboken, NJ, USA.

Different coating processes can result in a variety of coating thicknesses, morphologies, and electrical performances of the ICP coated textile fiber or yarn. Moreover, as fibers are converted to yarns and fabrics, they experience various stresses and strains that are unique to textile processing techniques. Additionally, if they are converted into sensors for commercial use, they would again be subjected to consumer use, such as laundering, shear stresses, and friction forces. Moreover, the coating process itself can add additional stress and can influence the mechanical properties of the fiber. Therefore, the harshness of catalysts, oxidants, and additives used during coating is important to consider. All these factors have to be taken into account while coating fibers with ICPs. As research in the field of ICPs grows and many interesting properties are achieved, there will be a need to consider the various requirements that textiles have and to ensure that a stable coated e-textile can be created [46].

2.2. Carbon Based Coatings

While the last section talked about a special class of polymers that are electrically conductive, most commonly used polymers are considered insulating materials. Within the area of e-textile sensors, various applications can subject the fiber sensor to large amounts of external stresses, which ICP coatings may not be able to withstand [58]. Hence, it is important to be able to convert commonly used polymers into robust electrically conductive systems. This can be done by adding electrically conductive filler materials, such as metal powders, fibers, carbon black (CB), carbon nanotubes (CNTs), and graphene, to the polymer matrix [59,60]. The mechanism of electrical conduction of such a system is based on percolation theory, which correlates the volume of filler material with the amount needed to create an electrically conductive pathway within the polymer matrix [61]. The first significant decrease in the electrical resistivity of the polymer composite is attributed to the formation of infinitely long contact chains between the conducting particles. The volume fraction of the conducting particles within the insulating matrix at which this decrease is first observed is known as the percolation threshold of that matrix and filler system [59]. As the volume fraction of conducting particles is increased beyond the percolation threshold, the resistivity continually decreases up to a certain volume fraction, beyond which continued addition does not cause a large decrease in the resistivity of the composite [61]. Hence, the percolation threshold is the lowest amount of filler concentration required within the polymer matrix to form an electrically conductive pathway [59]. There are a variety of factors that can be attributed to the conduction of electrons through a conductive polymer matrix—percolation theory, quantum mechanical tunneling, and thermal expansion [59]. Tunneling is significant at low temperatures and occurs when a high enough electric field is applied to excite electrons such that they can jump through the potential barrier—making resistivity a function of temperature and voltage. Thermal expansion occurs at high temperatures and causes the polymer matrix to swell, thereby increasing the distance between the particles within the matrix and resulting in an increase in resistivity. To realize conductive fibers and yarns for sensing applications, it is important to apply a composite coating that has good electrical performance as well as stability.

2.2.1. Carbon Nanotubes

CNTs have proven to be excellent candidates for the preparation of conducting polymer composites due to their unique structure coupled with excellent physical and chemical properties as well as mechanical strength [62]. CNTs are cylindrical structures composed of one (single walled CNT or SWNTs) or multiple (multiwalled CNT or MWNTs) layers of graphene with either open or closed ends [63]. With diameters ranging from 0.8–2 nm (SWNTs) and 5–20 nm (MWNTs) and lengths from <100 nm to several centimeters, CNTs can exist at multiple length scales [60], as shown in Figure 6. With the ability to achieve bulk conductivities of approximately 10^5 S/cm, CNTs are excellent materials to form conductive polymer composites, as their conductivity is within the polymer matrix following the percolation theory as described previously in this section [64]. In most cases, the percolation threshold for CNT-polymer composites is approximately 0.1 wt.%, with percolation thresholds below 0.1 wt.% attributed to systems with kinetic thresholds, where the filler particles can move and reposition themselves due to effects such as convection, diffusion, shearing, or externally applied fields [65]. This enables them to be good candidates for flexible coatings, since they do not require large amounts of filler to be electrically conductive [66]. Moreover, since CNTs themselves possess a high aspect ratio [60], this also has an effect on the percolation behavior of these composites. For statistical percolation, where the movement of filler particles themselves is ignored, the percolation threshold has been observed to decrease with increasing CNT length [67], whereas the electrical conductivity is highest when the CNTs are partially or slightly aligned rather than perfectly aligned and isotropic within the polymer matrix [68].

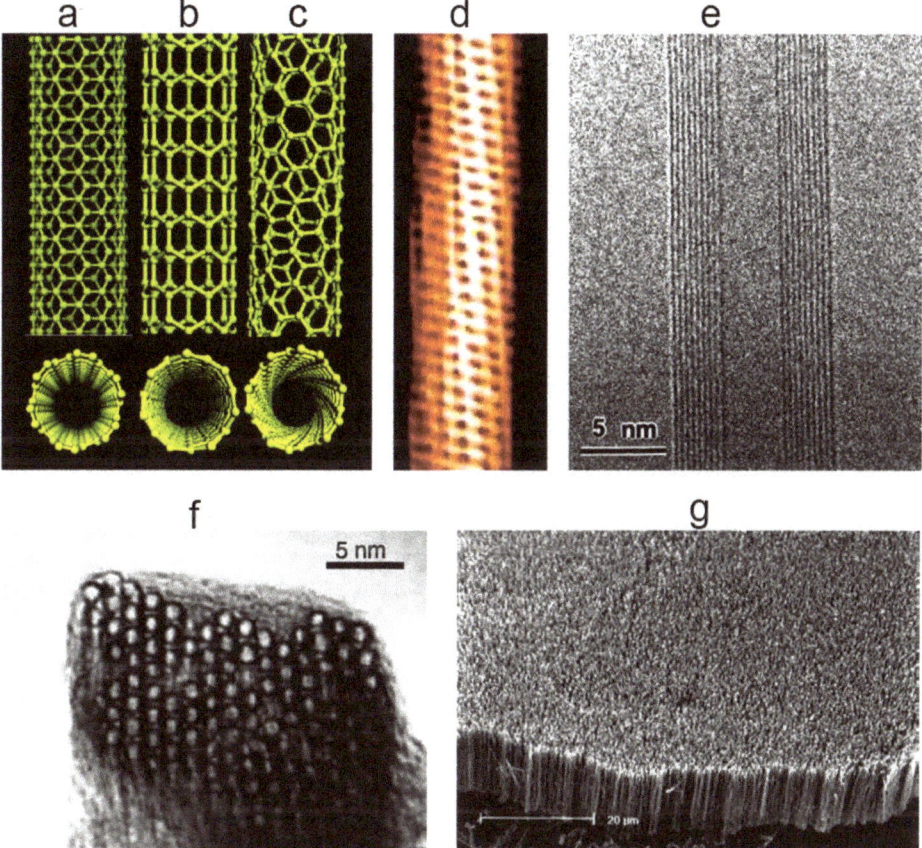

Figure 6. Schematic showing the (**a**) armchair, (**b**) the zigzag, and (**c**) the chiral single walled CNT (SWNT) structures. (**d**) Tunneling electron microscopy (TEM) image of helical 1.3 nm diameter chiral SWNT; (**e**) TEM image of multiwalled CNT (MWNT) with nine SWNTs concentrically nested in the structure; (**f**) TEM of 1.4 nm diameter SWNT in a bundle, shown laterally; (**g**) SEM of MWNT array grown as a forest. Figures (**a**–**c**,**e**,**g**) reprinted with permission from [69]. Copyright 2002 The American Association for the Advancement of Science: Washington, DC, USA. Figure (**d**) reprinted with permission from [70]. Copyright 1998 Springer Nature: Basingstoke Hampshire, UK. Figure (**f**) reprinted with permission from [71]. Copyright 1996 American Association for the Advancement of Science: Washington, DC, USA.

While CNTs themselves have been using in yarn form for strain sensing due to the observed increase in electrical resistance of CNT yarns with applied tensile strain [72], researchers such as Han et al. have integrated CNTs onto yarns by forming conformal coatings using CNT inks [58]. CNT inks were prepared by dispersing SWNTs into deionized water using sodium dodecylbenzene sulfonate (SDBS) as a surfactant. This dispersion was then coated onto cotton yarns using a paint brush, followed by air drying and washing to remove residual surfactant. Due to the large van der Waals forces between the CNT and the cellulose, as well as the multiple hydrogen bonds formed between the hydroxyl groups in the cellulose and the carboxyl and the hydroxyl end groups within the CNTs, the coating was able to percolate within the cotton yarns, resulting in a resistance of 7.8 kΩ/cm. These yarns were then used as ammonia sensors with a detection limit of 8 ppm and sensitivity ranging from 0–3% depending on the amount of time the sensor yarns were exposed to ammonia [58]. Shim et al. also developed cotton yarns with SWNT and MWNT coatings for use as e-textile biosensors, using the general dip-coating method to coat individual yarns with a dispersion of CNTs in Nafion solution [35]. They observed that the SWNT coated cotton yarns were more conductive (25 Ω/cm) than those coated with MWNT (118 Ω/cm), attributed to the tighter, more dispersed network formed by SWNTs compared to the larger and more rigid network formed by MWNTs on the same cotton substrate. Moreover, they also observed that post processing treatments such as thermal annealing could further reduce the resistance to 19 Ω/cm for the SWNT coated cotton yarns [35].

While these works involve coating CNTs directly onto fibers from dispersion, researchers have also prepared composites containing CNTs that are then coated onto textile yarns. Zhang et al. coated spandex multifilament yarns with a thermoplastic polyurethane (TPU)/CNT coating to form conductive elastic yarns that could act as strain sensors [66], as shown in Figure 7a–e. The yarn was driven over rollers into a TPU/MWNT composite bath where it was coated, followed by hot air drying, and finally coated yarns were collected on a rotating drum (Figure 7a). This process enables coating longer lengths of yarns continuously; however, there were some irregularities in the coating due to agglomeration of CNTs at certain locations on the yarn surface. Moreover, once a continuous coating was formed (at concentrations of CNT greater than 5 wt.%), there was no observable change in the resistance of the yarns as the concentration of CNTs increased. However, without a continuous coating, the yarns transitioned from semiconductor to insulator, thereby showing a dependence on coating thickness up to a certain CNT mass fraction. The resistivity of the CNT coating was obtained as 754 Ω·cm, attributed to the low viscosity of the coating solution enabling the formation of a kinetic percolation network. In another similar study, electrospinning, ultrasonication, and bobbin winding were combined by Li et al. to create SWNT/MWNT/TPU yarns (SMTYs), which were used as wearable strain sensors with the capability to sense large strains (100%), low hysteresis for 2000 cycles, and conductivity of 13 S/cm [73], as shown in Figure 7f–l. Electrospun TPU fibers were first collected in a water bath, where they floated on the surface of the water. From here, they were drawn across the water and bundled into a yarn structure via continuous rollers. These yarns then passed through U-tubes containing MWNT and SWNT dispersions in an ultrasonic bath that enabled the adsorption of the CNTs. Finally, the coated yarns were dried and collected on a take up roller (Figure 7f). A total content of 3.65 wt.% of combined MWNT and SWNT was applied on the yarn. The high surface area of electrospun TPU fibers ensured that a large amount of CNT could adhere to the surface. Additionally, the synergistic interaction between the adsorption of MWNTs on the surface of the TPU yarns followed by the conformal coating of SWNTs onto the MWNT-coated TPU yarn is attributed as the reason for high electrical conductivity.

While a lot of research is being performed on coating textile materials with CNTs [62,74], challenges still remain in terms of being able to form a continuous, uniformly disperse CNT coating solution as well as achieving good adhesion between the textile substrate and the CNTs [69]. Additionally, processes such as CVD are complex, high-cost, and restrict the lengths of CNT coatings that can be formed on fibers. Making composites with CNTs incorporated into them also limits the electrical conductivity of such a coating, since the polymer matrix is now covering the surface of CNTs. Additionally, CNTs have a tendency to aggregate within polymer matrices, thereby causing electrical and mechanical faults in the fiber sensors. Hence, other carbonaceous materials with lower costs and easier application techniques are also being explored as conductive coatings for e-textiles.

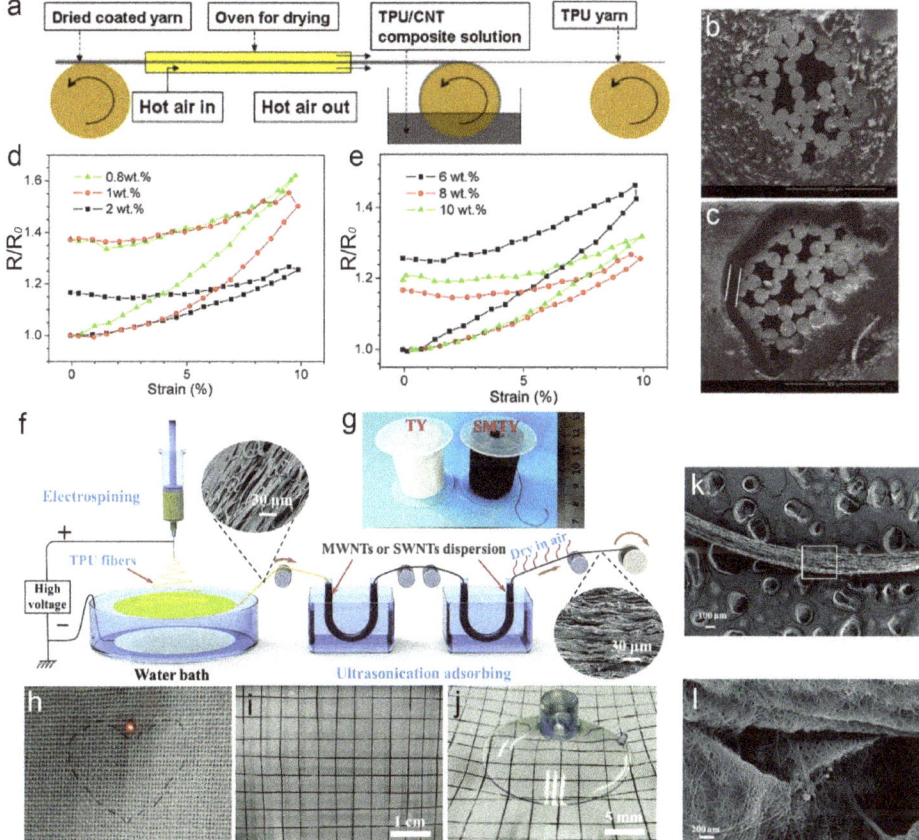

Figure 7. Thermoplastic polyurethane (TPU)/MWNT conductive polymer composite (CPC) coating for conductive, strain sensing yarns: (**a**) Schematic illustration of the coating process where the MWNT composite solution is applied onto TPU yarn; (**b**) cross-sectional SEM showing the uncoated TPU yarn. Scale: 300 μm; (**c**) cross-sectional SEM image showing coated yarn, with the two vertical white lines showing the exact coating location. Scale: 300 μm; (**d**) resistance-strain dependence on the amount of CNT concentration in the CPC; (**e**) resistance-strain dependence on the different levels of CPC applied onto the TPU yarn as coating. Figures (**a–e**) reprinted with permission from [66]. Copyright 2012 Elsevier B.V.: Amsterdam, The Netherlands. Formation of conductive SWNT/MWNT/TPU yarns (SMTYs) by combining electrospinning, ultrasonication, and bobbin winding: (**f**) Schematic illustration of the fabrication process for SMTYs; (**g**) uncoated (white) and coated (black) yarns obtained by the process; (**h**) electronic fabric lighting a light emitting diode (LED) using conductive SMTYs obtained in (**i**); (**i**) Optical image showing the electronic fabric composed of SMTYs; (**j**) SMTYs under bixaxial strain; (**k**) surface morphology of SMTYs shown in SEM with (**l**) showing higher magnification to indicate MWNT and SWNT coating. Figures (**f–l**) reprinted with permission from [73]. Copyright 2013 Royal Society of Chemistry: London, UK.

2.2.2. Graphene

Single-layer graphene is an excellent candidate for incorporation into e-textiles because of its high thermal conductivity at room temperature (~5000 W/m K), high Young's modulus (~1100 GPa), and charge carrier mobility (200,000 cm^2/V s) [75]. Graphene can be synthesized more cheaply than CNTs, does not require helicity control and has higher aspect ratio than CNTs, enabling it to be a cost-effective method of incorporating sensing functionality into textiles [76–78].

Graphite is a soft, black material with sp^2-hybridized carbon atoms stacked in two dimensional layers, with the layers themselves held together by van der Waals forces. An atomically thin layer of graphite forms graphene, which was first isolated in 2004 using a simple scotch tape [79]. Graphene oxide (GO) and reduced graphene oxide (rGO) derivatives or graphite are synthesized via solution-based oxidazition (Hummers method [80]) and reduction or via dry methods such as CVD [77]. Reduction of GO to form rGO is an important process usually done via exfoliation of the GO sheets using hydrazine hydrate. This is because GO is electrically insulating due to the loss of electrical conjugation caused by oxidative treatment during its conversion from graphite to GO [81]. Moreover, since GO consists of oxidized graphene sheets, it is hydrophilic and hence thermally unstable, as it can undergo pyrolysis at elevated temperatures. Chemical reduction to rGO results in the removal of oxygen and the reformation of unsaturated, sp^2 hybridized carbon sites, restoring the electrical conductivity of the graphene sheets [81].

Since aqueous solutions of graphene are easy to produce without the need for surfactants due to the presence of carboxylic and hydroxyl groups, it is a suitable material for coating onto textiles to impart electrical conductivity. Integration of graphene into textile fibers or yarns can occur via two methods: (i) mixing it with a fiber forming compound and then spinning to form compound fibers, or (ii) coating it onto already formed fibers or yarns made of nylon, cotton, or polyester [82,83]. Coating methods usually involve a dip and dry process; however, other methods of coating such as spraying have also been used to coat graphene on fibers [84]. Zhang et al. used a simple Meyer rod for a dry coating process that was repeated 10 times to press graphite powder onto silk, PP, and spandex fibers with further encapsulation by polydimethylsiloxane (PDMS) to form single fiber strain sensors [85]. This formed a conformal, core-sheath graphite-silk fiber with the coating attached to the fiber due to electrostatic and van der Waals forces. Li et al. also developed a core-sheath yarn structure with a graphene/poly(vinyl alcohol) (PVA) as the coating or the sheath material and polyurethane (PU) multifilament yarn as the core material [78]. This was done via a layer by layer (lbl) assembly method wherein the PU yarn was first coated with a PVA solution and then washed and rinsed such that the yarn could pick up 0.5 wt.% of PVA. This PVA acted as an adhesive for the graphene dispersion to be deposited onto the coated yarns. These two steps were repeated sequentially multiple times to create a thicker coating on the yarn, with the thickness varying from 523.2–2929.4 nm depending on the number of coating cycles and the concentration of graphene used. These two factors also determined the surface roughness of the coating and the subsequent lowering of resistivity as the number of coating layers and the amount of graphene coated onto the yarn increased. Hence, the performance of yarn sensors can be easily modulated by changing the thickness and the concentration of graphene. Yun et al. used an interesting method of applying rGO onto nylon-6 yarns using electrostatic self-assembly with bovine serum albumin (BSA) as an adhesive [86]. Since BSA is amphoteric, it can be attached to both hydrophobic and hydrophilic substrates and plays an important role in maintaining the structure of the textile during GO coating. Briefly, electrospun nylon yarns were functionalized with BSA via dip coating, dried, and then GO was assembled onto the surface by dipping these into a GO dispersion. The GO deposited onto the yarns was then reduced to rGO using a vapor reduction method with hydroiodic acid (HI) at low temperature. This process was also compatible with cotton and polyester yarn and could be adapted to fabrics, resulting in a nonwoven fabric with conductivity of 1040 S/m. Park et al. incorporated graphene nanoplatelets (GNPs) via lbl assembly as a conductive coating on three different types of stretchable yarns—rubber (RY), nylon covered rubber (NCRY), and wool yarns (WY)—for use as stretchable and wearable strain sensors [82], as shown in Figure 8. This was done by first coating the yarns with PVA, which attached to the yarns via noncovalent interactions. This was followed by dipping the yarns in a GNP solution containing poly(4-styrenesulfonic acid) (PSS) to form the conformal conductive coating (Figure 8a). The addition of PSS is crucial since it adsorbs to the graphene surface because of hydrophobic and π–π interactions and prevents agglomeration of the hydrophobic graphene in water due to the presence of hydrophilic sulfonic groups in the PSS structure. Moreover, van der Waals interactions and hydrophobic interactions enable binding of GNP with PSS to PVA on the yarns, thereby ensuring that a coating can be formed. The performance of these sensors is explored in Section 3 in greater detail.

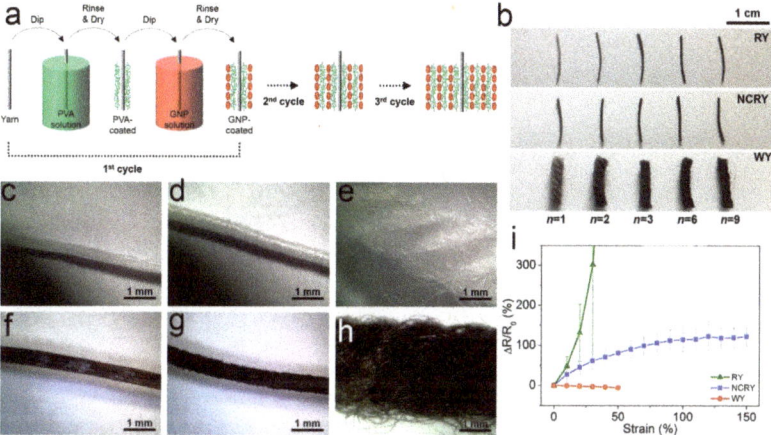

Figure 8. (**a**) Fabrication via layer by layer (lbl) method to coat rubber yarns (RY), nylon covered rubber yarns (NCRY), and wool yarns (WY) with graphene nanoplatelet (GNP) solution. The first step involves coating with poly(vinyl alcohol) (PVA) to enable adhesion of GNP to the yarns. By repeating the coating cycles, the thickness of the GNPs on the yarns can be increased; (**b**) image comparing the yarns after various numbers of coating cycles (n); (**c–e**) RY, NCRY, and WY before and (**f–h**) after three cycles of lbl coating; (**i**) relative percentage change in resistance as a function of strain showing the difference in piezoresistive performances between the various yarns. Reprinted with permission from [82]. Copyright 2015 American Chemical Society: Washington, DC, USA.

While graphene as a coating material in either rGO form or as a composite offers several advantages, research is still needed to produce rGO with uniform properties, especially for large-scale use in large-area fiber sensors. Moreover, it is important that the interface between the polymeric material and the graphene coating is robust, which may not always be the case for graphene-based coatings. Nevertheless, with its low cost and excellent electrical properties, graphene remains a promising coating material for e-textile sensors [77].

2.2.3. Other Carbonaceous Materials

While CNTs and graphene have dominated recent research in e-textile sensors, CB [87] and vapor-grown carbon nanofibers (VGCNFs) [88] have also been used for various coating applications to render textiles electrically conductive. Due to the low cost of these materials, they can be used as alternative materials to more expensive fillers such as CNTs [89]. CB is an amorphous form of carbon formed by the incomplete combustion of aromatic hydrocarbons at high temperatures. As an electrically conductive filler, CB should have a large surface area, moderate agglomerate size, and low volatile content to ensure that the polymer-CB filler has moderate melt viscosity as well as high electrical conductivity without requiring a high volume of CB loading [65]. Since CBs are usually large structured agglomerates, it is easier to apply them as coatings onto fabrics instead of fibers, thereby rendering electrically conductive e-textiles [90].

In terms of applications for fibers and yarns, CB is used more often as a conductive filler to spin or cast fibers using techniques such as melt spinning [91] or 3D printing [1]. Even so, in conjunction with other materials such as ICPs, CB has been used to apply conductive coatings onto yarns or fibers, such as the work done by Villanueva et al. to coat cotton yarn with a CB/PPy dispersion via dip-coating [92].

Using a PPy composite with 20% CB filler ensured that the conductivity of the ICP could be enhanced inexpensively while still being able to coat it onto cotton yarns. By repeating the coating and drying process 15 times, they were able to create conductive cotton yarns with electrical conductivity of 12.6 S/m as well as a reduction in resistance from 9 MΩ for untreated yarns to 129 kΩ for the coated ones. Souri et al. used natural flax and flax bleached (FB) yarns as biodegradable substrates for coating with CB and CB-graphene nanoplatelets (GNPs) using a novel ultrasonication process to fabricate electrically conductive flax yarns for use as a pressure sensor [93]. Using a deionized (DI) water suspension of CB and CB-GNPs with SDBS as the surfactant, they were able to ultrasonically coat flax and FB yarns with the conductive dispersion. These yarns were then sandwiched between PDMS layers to form the final yarn sensor. With their large sizes (30 nm), CB particles formed conductive bridges between the GNPs (5 nm thickness, 10 μm width), thereby enhancing the electrical conductivity of the yarn to 585 S/m with just 20 min of ultrasonic coating. Hence, CB particles are good fillers to enhance the already present electrical conductivity of various other fillers, such as GNPs, graphene, and CNTs.

CNFs are produced by catalytic CVD of carbon monoxide or a hydrocarbon over a metal or a metal alloy catalyst with the reaction proceeding at 500–1500 °C [89]. VGCNFs are similar to MWNTs in structure with a hollow nanofiber core composed of single or double graphene layers stacked parallelly or at an angle to the fiber axis. They have high aspect ratios with diameters ranging from 15–200 nm and lengths in the range of tens of micrometers [94]. In polymer composites, the percolation threshold of VGCNFs varies from 9–18 wt.% for PP/VGCNFs, 3% for polyacrylonitrile (PAN)/VGCNFs, and 6% for PC/VGCNFs, with the added enhancement of Young's modulus of the polymer due to it being a fiber-shaped composite filler [95]. As with CBs, VGCNFs have been used mostly to apply conductive coatings onto fabrics rather than yarns or fibers, which is to be expected from larger sized filler materials that are harder to apply to fiber or yarns [94,96]. They have also been used to form electrically conductive composites that can be used as flexible sensors [88,97]. Nevertheless, Narayan et al. used 5 and 10 wt.% of CNF in a thermoplastic polyurethane (TPU) matrix to form a coating material that was subsequently applied to cotton yarn and silk filaments via dip-coating [98]. Since cotton yarns have a more expanded structure with higher porosity than silk filaments, the coating solution penetrated deeper into the cotton yarns, whereas it remained more on the surface in the case of silk, resulting in an increase in diameter for the latter. Additionally, fillers with high aspect ratios, such as CNTs and CNFs, showed a marked improvement in the electrical conductivity of the yarns. The CNF dispersed nanocomposite coating showed a resistance of <100 kΩ/cm compared to a much higher resistance (10^9 Ω) for uncoated cotton and silk fibers. Silk is more conductive when coated, since more of the coating can adhere together onto the surface rather than penetrate deeper into the voids and lose percolation, as it happens in the case of cotton fibers. Hence, they were able to develop conductive cotton and silk yarns with maximum bulk conductivities of 2 S/m and 12.5 S/m, respectively.

Compared to graphene and CNTs, CNFs are more suitable for fabric-based coating applications and can be applied onto fabrics in numerous ways, such as: (i) direct growth of CNFs onto the fabric substrate via CVD [99], (ii) deposition of CNF onto the fabric layer [99], and (iii) electrophoretic deposition of CNF onto the fabric substrate [100].

2.3. Metals

While carbonaceous coatings are flexible, lightweight, and can be easily applied to the fiber or yarn structure, they fail to provide excellent electrical conductivity, low contact resistance, and simultaneous structural stability [24]. For this reason, metals are coated onto fibers or yarns to impart electrical conductivity. Electroplating (also known as electrodeposition) [101] and electroless plating [102] are two very common and versatile methods of metallization of polymeric materials such as textiles. However, other techniques such as chemical solution processing have also been used to impart metallic coatings onto textiles [103]. Electroplating refers to the process of film growth on a substrate through the electrochemical reduction of metal ions from the electrolyte [104]. Various factors affect the coating during electroplating, such as the substrate and coating interface, the coating material itself, and the coating-environment interface [105]. While regular electroplating techniques involve the use of an externally applied electric current to drive the metal displacement reaction, electroless plating builds metallic deposits by chemical reactions without consuming the substrate material. This is done by the selective reduction of metal ions only at the surface of a catalytic substrate, which is immersed in an aqueous solution of the same metal ions, resulting in continuous deposition on the substrate through the catalysis of the deposit itself [106]. Hence, electroless plating is also referred to as autocatalytic plating, as the deposit catalyzes the reduction reaction.

Depositing metals on yarns involves chemical transformations taking place on the yarn surface, which should be free from impurities [101]. Little et al. compared electrochemical deposition and electroless plating of nickel (Ni) and gold (Au) on Kevlar fibers to render them electrically conductive [101]. Since Kevlar does not have surface chemical groups that can bind onto metal ions, they were pretreated with tin (Sn) ions followed by deposition of palladium (Pd) particles. Pd was believed to act as a catalyst for electroless deposition of nickel using hypophosphite ions, which were formed on the Kevlar fiber using electroless deposition. However, this method failed to create metallic Ni coated, instead creating an amorphous coating believed to be amorphous phosphides that crystallize at high temperatures to form Ni_2P, Ni_3P, and $Ni_{12}P_5$. These Ni treated yarns were used as starting substrates for electroless deposition of Au. However, this technique of electroless Au deposition was not very reproducible or robust. On the other hand, electrochemical deposition of Au on Ni-treated (electroless) Kevlar fibers produced a more uniform coating with lower resistance, better reproducibility and adhesion, and higher mass gains of the coating [101]. Hence, it is important to choose the right surface pretreatment and type of metal being deposited during electroless deposition to ensure that a conductive, conformal, and uniform metallic coating is produced on the fiber.

Liu et al. produced a polyelectrolyte bridged copper (Cu) coating on cotton yarns to produce conductive (1 S/cm) yarns in a process that could also be used for making other natural fibers and fabrics conductive [102]. In this process, d-poly [2-(methacryloyloxy)ethyltrimethylammonium chloride] (PMETAC) brushes were synthesized from cotton yarns using atomic transfer radical polymerization (ATRP). ATRP was used because it enables fast polymerization with good control over thickness, density, and uniformity of the PMETAC brushes formed. This process was followed by immersion of these modified cotton yarns into an aqueous Pd solution since Pd is a good catalyst for the electroless deposition of Cu. $PdCl_4^{2-}$ species were immobilized onto the PMETAC brushes due to their high affinity to ammonium groups. Finally, the yarns were coated with Cu particles in an electroless plating bath with a 60-min coating, resulting in 1 S/cm conductivity of the yarns. The advantage of this method is that it produces a connecting bridge between the Cu and the cotton substrates that imparts robustness to the coating during bending and stretching. Additionally, during stretching, the fibers within the yarn pack more closely than when they are in the relaxed state, thereby exhibiting higher conductivity when stretched (0.28 S/cm) than when they are relaxed (0.04 S/cm) for a Cu coating applied for 30 min. However, Cu coatings do not exhibit good air stability, with a 10% decrease in electrical conductivity after a seven day exposure to air. To improve on this, Liu et al. coated the cotton yarn with Ni, which, while exhibiting a lower electrical conductivity of 0.3 S/cm, was still able to retain that performance over the course of two months. Moreover, no reduction in conductivity was observed during mechanical testing or washing [102].

Even though electro- and electroless deposition techniques are well-studied and useful for metallizing fibers, there are numerous problems that are encountered in these techniques. To form a uniform coating on longer lengths of yarns or fibers with metals such as Ni or silver (Ag), these techniques and the raw materials required can be quite expensive. Moreover, poor electrical conductivities and mechanical performances are encountered in fibers made conductive via electroless plating [103]. Hence, other techniques to metallize textile fibers and yarns have also been explored.

Lee et al. used an alternative method to coat aluminum (Al)—a more cost-effective metal—onto cotton thread using a chemical solution (CS) process [103]. By pretreating the cotton fibers with a fumed catalyst, titaniumisopropoxide, followed by immersion in the Al precursor composite solution,

Al was able to easily penetrate into the fibers. Moreover, nucleation of Al occurred on the surface of the fibers at room temperature, growing large enough to cover all the fibers. Hence, in this way, there was a surface coating as well as deeper penetration of the metal into the cotton fibers. Compared to commercially metallized cotton yarns (1000 Ω/10 cm), these yarns had a much lower resistance (<30 Ω/10 cm) due to the dense structure formation of Al on the fibers. Jur et al. also used atomic layer deposition (ALD)—a vapor phase method—to produce thin films of silver on nylon fabric and zinc oxide (ZnO) on woven cotton fabrics and nonwoven PP [107]. ALD involves coating a surface by exposing it sequentially to a metal organic precursor followed by a reactant. This results in the formation of a complementary sequence of self-limiting reactions, which can occur at temperatures less than 150 °C, thereby making this a compatible way to deposit oxides, nitrides, and conducting films on conventional textile substrates. Park et al. developed Ag-Au nanoparticles (NPs) coated cotton yarns via a simple dip-coating technique followed by electroless Ag deposition to create flexible strain sensors that could monitor human motion [108]. The cotton yarns were functionalized with amine groups, and AuNPs were then bonded onto the cotton yarn surface. The amine groups created a positive charge on the yarn surface and on immersion in an aqueous Au NP solution, and a uniform coating on negatively charged Au NPs could be formed on the yarn surface. This was followed by immersion of these Au-cotton yarns in an aqueous Ag solution to obtain the final Ag-Au cotton yarns. The positively charged Ag ions were able to attach onto the Ag-cotton yarn surface through electrostatic interactions, forming thin Ag shells on the surface of the yarn via the reduction of the Ag ions with hydroquinone. With a minimum resistance of 90 Ω/cm, these yarns were suitable for strain sensing applications. Moreover, when bent more than 9°, these yarns showed an increase in electrical conductivity, which was attributed to the individual conducting microfibers attaching to the center of the yarn, thereby forming a much more conductive pathway due to increased contact between the microfibers. When bent less than 9°, their resistance increased due to the decrease in contact area between the individual microfibers. Hence, these sensors could register fine movements such as finger bending with high sensitivity [gauge factor (*GF*) of 20 for strains <9°] [108]. Section 3.2 explores performance parameters such as sensitivity and *GF* in more detail. Lee et al. also developed fiber-based pressure sensors (based on a capacitive sensing principle that is further discussed in Section 3) wherein the conductive fibers were fabricated by coating poly(styrene-block-butadiene-styrene) (SBS) rubber on the surface of Kevlar fibers and subsequent conversion of Ag ions to Ag NPs directly on the surface of the SBS polymer [24]. This was done in three steps: (i) coating SBS on the Kevlar fiber by essentially flowing the SBS down the Kevlar fiber, (ii) adsorption of Ag onto the SBS layer, which was achieved by immersing the SBS coated Kevlar fiber in AgCF$_3$COO solution and ethanol, followed by (iii) reduction of the Ag precursors formed in the previous step to fabricate Ag NPs on the SBS layer using a solution of hydrazine hydrate. Densely coated Ag NPs with diameters of 70–90 nm were thus formed with good connections to each other to render the Kevlar fibers electrically conductive (resistance of 0.15 Ω/cm) and comparable to commercial threads (1 Ω/cm) and those made by Lee via the CS method (0.2 Ω/cm) [103]. Their sensing parameters are further discussed in Section 3.4.2.

While metal coatings have their own advantages, they also present certain limitations in terms of durability of the coatings themselves, since they present a transition at the interface from a soft fiber to a rigid metallic coating. Moreover, many times, the techniques used to coat metals can be expensive, along with the metals used themselves [103]. Metallization techniques also suffer from being unreliable in terms of creating a uniform, thick, and robust coating. Therefore, various novel techniques are being explored, as discussed in the previous paragraph.

Research on fiber-based e-textile sensors is based on the premise that textiles provide a unique avenue to develop wearable sensors and that the fiber level integration may be the most appropriate of all. This approach also opens traditional textile processes, such as yarn formation, weaving, and knitting, to produce textile structures for capacitive and resistive sensing modalities. The subsequent sections discuss basic sensing principles and metrics of various fiber-based sensors that were made using the coating methods and the materials discussed in Section 2. Additionally, since strain and pressure are two very well studied and highly researched sensing modalities in e-textiles [24], the discussion in Section 3 is limited to these two sensor types.

3. Coated Fiber-Based Pressure and Strain Sensors

Flexible fiber-based sensors have been suggested for monitoring various stimuli, including temperature [109], humidity [110], chemical levels [111], pressure [112,113], and strain [66,73]. Fiber-based sensing is advantageous when compared to film based sensors because fibers are lightweight materials with high aspect ratios. This allows fibers to easily conform to the contours of the human body, such as wrists and fingers—locations where wearable technology has garnished great attention [114,115]. Further, fiber-based sensors may be incorporated into various textile structures, including woven and knitted materials, allowing for flexibility of design [48]. The fiber form-factor may be the most promising building block for future wearable technology, including sensors, actuators, and artificial muscles [72]. The most widely researched fiber-based sensors are employed for monitoring strain and pressure [114,116–119]. Fiber-based pressure and strain sensors have been widely researched, as their fabrication and sensing mechanisms are relatively simple. This section focuses on the fiber-based strain and the pressure sensors fabricated via coating methods, their sensing mechanisms, and performance. Additionally, proposed applications for such technology are discussed.

3.1. Sensing Principles

Strain and pressure sensors convert mechanical stimuli (compression, bending strain, flexion, twisting, etc.) into electrical signals that can then be monitored. Pressure and strain sensing can be achieved by various principles, including optical [120–123], piezoelectric [124], hybrid piezoelectric and triboelectric [125], resistive [116,126], and capacitive sensing [24,26,127]. Resistive and capacitive type sensors are most often employed due to their facile fabrication, ease of use, and relatively simple electronics [128,129]. Hence, the focus of this review is on resistive and capacitive sensing modalities used for monitoring strains and pressure.

3.1.1. Piezoresistive

Piezoresistive sensing of pressure and strain is demonstrated schematically in Figure 9a. Such sensors utilize materials that undergo a change in electrical resistance when subjected to an external deformation [126,129]. Piezoresistive sensors require a single electrode that serves as a resistor. A voltage is applied to the resistor, and changes in resistance upon deformation can thus be monitored. The electrical resistance of the material varies according to the following equation:

$$R = \frac{\rho l}{A} \tag{1}$$

where ρ denotes electrical resistivity, l is the length of a sample, and A is the cross-sectional area of the resistor. As shown in Equation (1), changes in resistance may be due to geometrical changes in the resistor's area (A) or length (l). Alternatively, changes in resistance may be due to changes in a material's resistivity. Electrical resistivity (ρ) is an intrinsic material property, and for homogeneous materials, resistivity is invariant. Therefore, the piezoresistive behavior for homogeneous materials is derived from the changes in resistor geometry (A, and l). However, for bi-phasic systems such as polymer composites, changes in resistance may also be attributed to composite materials' changes in resistivity. The strain-induced change in resistance is known as piezoresistive behavior. Piezoresistive materials are those that change resistivity upon deformation [117,126,130]. At an atomic level, piezoresistance may be explained by changes in energy gaps between valence and conduction bands, which alter the number of charge carriers, ultimately changing a material's resistance [131]. Piezoresistive behavior of fiber-based sensors is often attributed to improvement or disruptions of electrical pathways within a conductive network of conductive polymers or particles upon deformation [129]. When a mechanical stimulus is applied to such fiber-based sensors, an increase (positive piezoresistance [116,126,130,132]) in resistance due to the disruption of conductive pathways may be observed. Alternatively, a decrease (negative piezoresistance [52,82,114,133]) in resistance may be observed due to the formation of new electrical pathways, which improves conductivity. Examples of resistive fiber-based sensor devices are shown in Figure 9b–d, and their electrical responses are shown in Figure 9e–j.

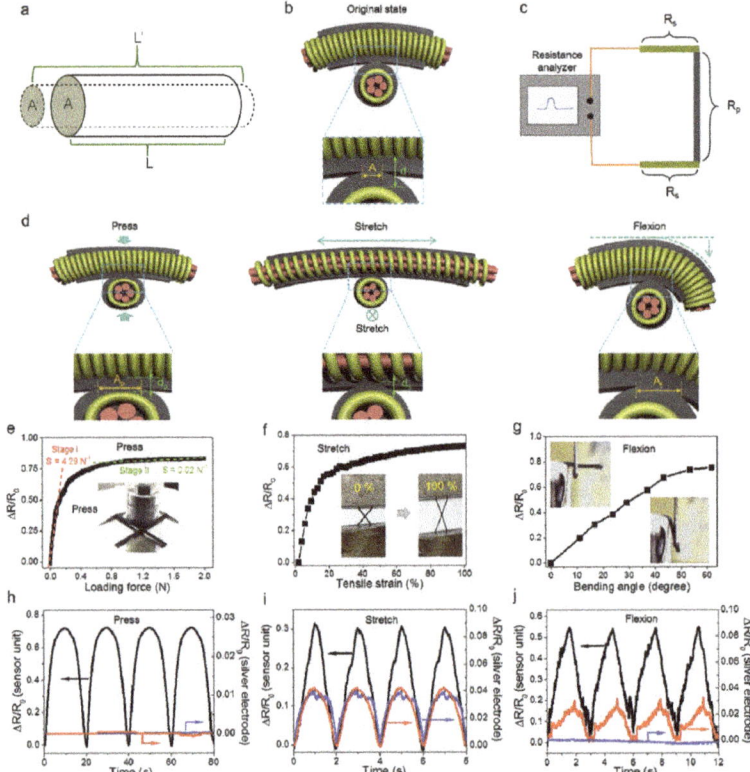

Figure 9. Piezoresistive sensing modalities: (**a**) Schematic representation of piezoresistive sensing; (**b**) schematic of cross-over contact point of two fiber sensors. "A" represents contact area and "d" indicates the separation between conductive electrode layers; (**c**) equivalent circuit for fiber sensor; (**d**) shape deformation at the contact point of the sensing unit when pressed, stretched, or flexed. A_p and d_p are the contact area and the thickness when pressed, d_s is the thickness when stretched, and A_f is the contact area during flexion, respectively. Plots of relative change in resistance ($\Delta R/R_0$) under (**e**) compression; (**f**) tensile strain; and (**g**) bending. Relative change in resistance of the sensor unit (black curves) and silver electrodes (blue and red curves) as a function of time at incrementally increasing and decreasing: (**h**) loading force (0–0.35 N); (**i**) tensile strain (0–8%) and (**j**) bending angle (0°–35°), respectively. Figures (**b–j**) reprinted with permission from [126]. Copyright 2015 WILEY-VCH Verlag GmbH & Co. KGaA, Weinheim: Weinheim, Germany.

The resistance response to deformation depends on the material make-up and the structure of the sensor as well as the mechanical stimuli being applied. In the case of strain sensors, it is more common that resistance increases with strain due to disruptions and breakages within the conductive network [82,85,115,118,126]. However, there are some cases in which a decrease in resistance may be observed with strain application due to improved electrical pathways [52,133,134]. Piezoresistive pressure sensors are less common when compared to capacitive mode pressure sensors. However, pressure sensors within the scope of this review note both increases [126] and decreases [114] in resistance with pressure application. In the case of piezoresistive pressure sensors, changes in resistance are often attributed to changes in contact area resistance between fibers arranged in a yarn configuration or fabric array [114,117,126].

3.1.2. Capacitive

Capacitive sensors convert mechanical stimuli to an electrical signal via a change in their capacitance [131]. Within this review, the use of capacitive sensors is restricted to pressure sensing. Capacitive sensors consist of a dielectric layer sandwiched between two parallel conductive surfaces (electrodes) (see Figure 10). A dielectric material is an electrical insulator. However, unlike pure insulators, when external electric fields are applied on dielectric materials, the electrons move only slightly away from their normal position. The slight displacement or movement of electrons is said to polarize the dielectric. The movement of the charges against restraining molecular forces provides the material with the ability to store electric energy. The parameter used to represent the relative (compared to perfect vacuum) charge storage capability of a dielectric material is the dielectric constant or relative permittivity (ε_r) [135]. When a direct current (DC) voltage is applied to a capacitor, charges accumulate on the two electrodes, while the dielectric layer prevents current flow [129]. Hence, a capacitive signal is generated, which is subsequently measured. The relationship between capacitance, C, the area of the conductive electrodes, A, the distance between the conductive electrodes, d, the permittivity of free space, ε_0, and ε_r is shown in Equation (2).

$$C = \frac{A\varepsilon_0\varepsilon_r}{d} \tag{2}$$

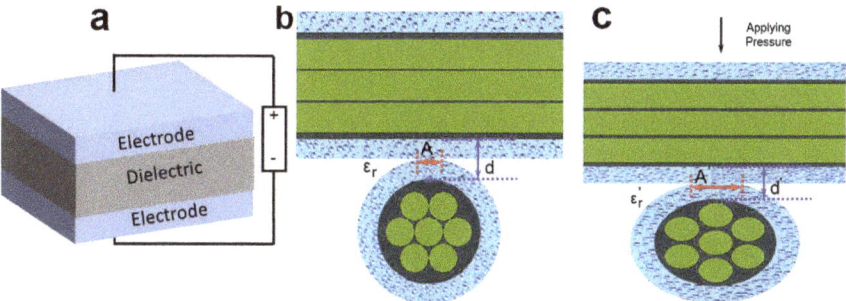

Figure 10. Schematic representation of capacitive pressure sensing: (**a**) Parallel-plate capacitor (electrodes shown in blue, dielectric in grey); (**b**) cross-sectional view of fiber-based capacitive sensing unit prior to deformation; (**c**) structural and dielectric changes induced by pressure application. In this diagram, A increases, d decreases, and the dielectric constant of the porous dielectric changes due to air removal with pore compression. Figures (**b**,**c**) reprinted with permission from [119]. Copyright 2013 Royal Society of Chemistry: London, UK.

Capacitive sensors measure the change in capacitance between the two conductive plates when an external stimulus is applied. If this stimulus causes a change in the overlap area between the electrodes (A), a change in the distance between the electrodes (d), or a change in the dielectric behavior of the material itself, a subsequent change in C can be measured [131]. Various capacitor geometries include parallel plate (shown in Figure 10) and cylindrical. Parallel plate capacitors are most common in fiber-based, coated sensors. In parallel plate capacitors, fibers or yarns are coated with a conductive material and then a dielectric material. When such fibers are incorporated into a yarn or fabric array, the fiber crossover points act as capacitors [24,118,119,136]. However, cylindrical configurations have also been explored for fiber-based sensors [2,112,137].

3.2. Performance Parameters

Gauge factor (GF) and sensitivity (S) are critical and commonly cited sensor performance parameters. GF and S indicate the ratio of change in sensor output to the change in the measurand [138]. GF indicates the relative change in electrical resistance, (R), in response to strain, (ε), whereas S indicates the relative change in capacitance, (C), or resistance, (R), with applied pressure, (P), as shown in Equations (3) and (4), respectively.

$$GF = \frac{\Delta R/R_0}{\varepsilon} \quad (3)$$

$$S = \frac{\Delta C/C_0}{P} \text{ or } S = \frac{\Delta R/R_0}{P} \quad (4)$$

It is desirable to maximize GF and S such that small changes in strain or pressure may be detected. One way to do so is by creating highly conductive sensors. This is also advantageous because highly conductive sensors consume less power to operate [114,139]. Therefore, conductivity/resistance of sensors is an important parameter to consider [24]. Another oft-reported parameter that indicates the maximum strain or pressure that can be measured is sensing range [138]. Range of sensing is often dictated by the material limits and should be optimized for a given application. In terms of pressure sensing, medium pressure regions (suitable for detecting object manipulation) ranges from 10–100 kPa, whereas pressures <10 kPa are comparable to gentle touch [140,141]. In terms of strain, the ability to detect the full range of human motion has been reported with ranges between 0.1–150% [132] as well as 0.2–100% strain [116]. Many researchers strive to achieve both large GF or S while also maintaining a large sensing range, which has proven to be quite challenging, as these parameters are often counteractive [114,116].

Sensors should be adequately durable from a mechanical standpoint to withstand cyclic testing and ideally provide stable electrical results throughout use. In terms of the wearable technology market, garments may be stretched repeatedly over their lifetime; therefore, it is important that such sensors are able to withstand cyclic testing. Researchers have tested pressure and strain sensors to a large range of cycle numbers to indicate mechanical durability and electrical stability. Cyclic tests ranging from 1000 cycles [52,116] to 100,000 cycles [26,126] have been reported to prove durability and stability. Stability is often indicated by the drift of response during cyclic testing. Drift is the amount that the electrical signal changes over the course of cycles.

Another critical sensor parameter is hysteresis. Hysteresis indicates the difference in the two output values during the increase or the decrease in the measurand. For example, in the case of pressure, sensing the difference in sensor output during loading (strain or pressure application) versus unloading (strain or pressure relaxation) is a measure of hysteresis [131]. A sensor should provide similar responses with minimum hysteresis when subjected to cyclic testing. It is unclear what degree of hysteresis is acceptable. However, hysteresis has been considered negligible if below 5% [127] or 6.3% [119]. Another commonly reported performance is stretchability and flexibility. Particularly in the field of wearable electronics, it is important that sensors are able to stretch and flex with the users. Conventional strain gauges provide a workable strain range of <5% [82,142], which is inadequate for wearable applications. During the basic movement of walking, the skin on the feet, the waist, and the joints are repeatedly stretched to as much as 55% [28]. However, it has proven challenging to achieve high conductivity, stretchability, and sensitivity [114]. Stretchability and flexibility of the fiber sensor relies on the concept that, when stretched or flexed, conductive networks are not destroyed [114].

Furthermore, strain and pressure sensors should be capable of detecting motions at a range of frequencies. The exact frequency desired for measurement is dependent on the measurand of interest. For example, step frequency at low speeds of running is approximately 2.5 Hz and increases curvilinearly with increasing speed [143,144]. Depending on the stimuli of interest, a sensor may need to be responsive at frequencies higher or lower. Related sensor performance parameters include response time and relax time, which indicate how quickly a sensor responds to an applied stimulus and how quickly a sensor returns to its baseline electrical signal after the stimuli is removed. When considering fiber based sensors, the response and the relaxation behavior will be highly dependent on the viscoelastic properties of the polymeric material [114]. It is unclear what response and relax times are desirable; however, papers report response times of <100 ms [114–116] and relax times of 10–15 ms as desirably fast [24,114].

3.3. Coatings for Fiber-Based Strain Sensors

While many papers exist that explore coating entire fabrics to create strain sensors [145–152], this section is limited to fiber and yarn based sensors. In most cases, strain sensors are proposed for measuring tensile strain. However, many papers also demonstrate the ability to sense bending and torsion [115,116,126] (see Figure 9c,g,j). Wearable strain sensors may be bent and twisted during practical use. Therefore, it is advantageous to study sensor behavior in these modes in addition to tensile deformation [115]. All strain sensors covered in this review sense by monitoring changes in resistance with mechanical deformation, showing either an increase or a decrease in resistance depending on the material properties and the sensor structure. Strain sensors in this review are primarily proposed for motion monitoring for applications in healthcare [153], virtual reality [115], electronic skins [118], and robotic systems [114]. However, a few researchers have studied the potential to utilize fiber-based strain sensors in composite monitoring for prevention of structural damage [154,155]. It is important to note that, in structural monitoring applications, strain measurement requirements are much lower compared to wearable applications.

3.3.1. Intrinsically Conducting Polymer Coated Strain Sensors

ICP structures and coating techniques were discussed earlier in Section 2.1. Here, fiber-based strain sensors that comprise ICPs are discussed. Eom et al. created a textile strain sensor proposed for wireless user interfaces using polyester fibers coated with PEDOT via in-situ polymerization followed by encapsulation with PMMA [52]. The fabrication method was previously discussed and is shown in Figure 4. The PEDOT/PET fiber provided electrical resistance of ~600 Ω/cm and a *GF* of approximately −0.76 at 20% strain, −0.665 at 50% strain, and −0.244 at 70% strain. When stretched, the resistance of the fiber decreased. This sensor's electromechanical behavior was attributed to the multifilament structure of the PEDOT/PET fiber. When stretched, the monofilaments became more interconnected, thus increasing fiber conductivity. The fiber sensor was tested over 1000 cycles (20% strain) during which a rise in resistance was observed; however, the *GF* remained stable. These sensors were then integrated into textiles via sewing with a commercial sewing machine and could be used to monitor body motion, as shown in Figure 11.

Fan et al. created a wearable strain sensor intended for smart clothing applications using PU fibers coated with PANI via in-situ chemical oxidative polymerization [130]. They reported fiber conductivity as 10^{-2} Ω/cm when using 6–7 wt% PANI. These fibers could sense strains up to 1500%, with the most sensitive region being <400% (*GF* of three). *GF*s of remaining strain ranges were not reported. Three regions of strain sensing behavior were observed. In the first region of strains (0–400%), sensing was attributed to disruptions in the conductive layer, which caused increases in resistance. In strain regions from ~400–1200% resistance changes with strain became less pronounced. This was attributed to greater PANI interconnection, which reduced the rate of resistance change with strain. Finally, in strains >~1200%, resistance continued to increase due to breakage of the electrical pathways within PANI. While these fibers did provide a large sensing range and high *GF*, their behavior was irreversible and highly hysteretic when exposed to cyclic testing at 50% strain. Additionally, fiber conductivity decreased significantly with washing, from 10^{-2} Ω/cm to 10^{-5} Ω/cm after washing for five minutes. Wu et al. created a PU-based fiber sensor with CNT and PEDOT:PSS coatings [132]. The PU fibers were first dip coated in a CNT dispersion and then dipped into a PEDOT:PSS hydrogel, followed by soaking in methanol to improve conductivity. Fibers were then twisted to enhance robustness and durability. Finally, the conductive fibers were coated in silicone, which served as a protective layer. Fibers were then prestrained to 50% to enable crack formation in the PEDOT:PSS layer. These fibers had a hierarchical structure in which the micro-cracked PEDOT:PSS sensing layer was connected by the conductive CNT agglomerates. The CNTs acted as conductive bridges, ensuring conductivity was maintained even at large strains. When the fiber was strained, the connection between the PEDOT:PSS fragments and the CNT agglomerates decreased, causing an increase in resistance. The authors report a high sensitivity (*GF* up to 350 for 150% strain), a wide sensing range (0.1–150% strain), low hysteresis, good linearity of response (up to 50% strain), good reliability (>2000 cycles at 50% strain), high cycling stability, and good repeatability. These fibers could sense the full range of human motions from subtle (pulse, phonation) to large movements (knee joints). Additionally, the fibers were able to detect strain distributions when configured into an array.

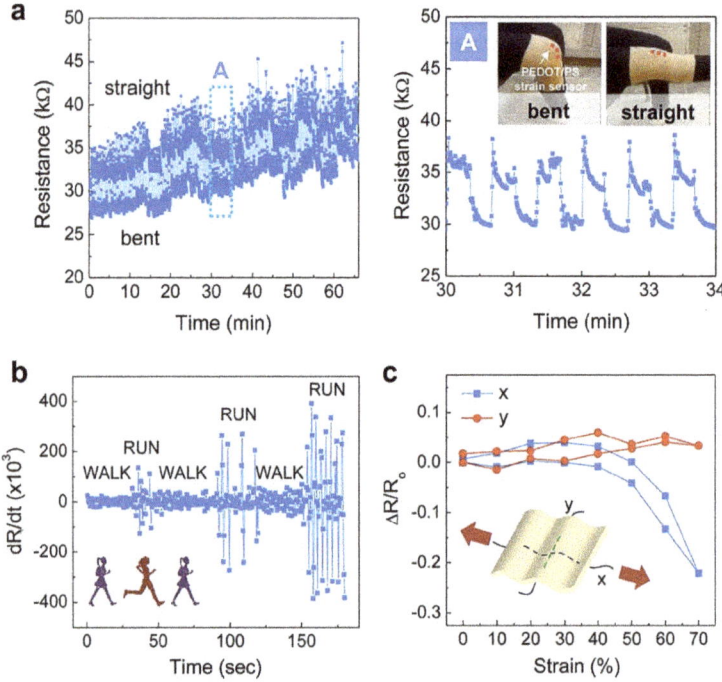

Figure 11. (**a**) Change in resistance of PEDOT coated PET textile while monitoring knee motion, with the top left figure showing good repeatability under multiple bending and straightening cycles of the knee (approximately 1000 cycles). The top right figure shows a zoomed in plot of the region "A" indicated in the top left graph; (**b**) demonstration of PEDOT coated PET textile sensor for monitoring walking and running motions; (**c**) PEDOT coated PET textile sensor monitoring strain in orthogonal directions (x and y axis) of fabric. Reprinted with permission from [52]. Copyright 2017, American Chemical Society: Washington, DC, USA.

Pertinent details related to fiber-based strain sensors coated with ICPs are provided in Table 1. These sensors may provide relatively good conductivity and high *GF*s when combined with CNTs [132]. However, some results indicate that strain sensor incorporating ICPs are hysteretic and not durable to washing [130]. While fiber-based sensors incorporating ICPs are appealing due to their fully-polymeric nature, ICPs are generally brittle materials that may be unable to withstand subsequent processing [58]. Therefore, conductive carbon based materials, such as those discussed in the following sections, have been explored.

Table 1. Fiber-based strain sensors coated with intrinsically conducting polymers.

Materials	Fabrication	Sensing Mechanism	Sensor Properties	Applications
PET yarns and PEDOT encapsulated in PMMA [52]	In-situ polymerization	Interconnectivity between the PET/PEDOT monofilaments enhanced on straining, decreasing resistance	• Resistance: ~600 Ω/cm • GF: −0.76 (20% strain), −0.665 (50% strain), −0.244 (70% strain) • Durability—withstood 1000 cycles with stable GF	Human motion monitoring
PU yarn and PANI [130]	In-situ chemical oxidative polymerization	Region 1: conductive pathways broken. Region 2: increased contact between PANI chains, reduced rate of resistance increase. Region 3: PANI conductive pathways broken, resistance increased	• Conductivity: 10^{-2} Ω/cm • GF: ~3 (400% strain) • Sense up to 1500% strain • Reversible response • Non-repeatable cyclic response	Smart clothing
PU yarn and PEDOT:PSS, CNTs [132]	Dip-coating	When strained, the number of electrical connections between PEDOT:PSS fragments and CNTs agglomerates decreased, resistance increased	• GF: 350 (150% strain) • Sensing range: 0.1–150% strain • Durability: withstood 2000 cycles (50% strain) • Linearity up to 50% strain • Negligible hysteresis	Human motion detection of subtle and large deformations

GF: gauge factor, PSS: poly(4-styrenesulfonic acid), PU: polyurethane.

3.3.2. Carbon Coated Strain Sensors

Fiber based strain sensors have been made through various coating techniques and with different carbonaceous particles. While carbon nanotubes [48,112,133,153] are most commonly employed to impart electrical capabilities, graphene [82,116] and CB [156] have also been explored.

CNT Coated Strain Sensors

CNT structures and coating techniques were previously discussed in Section 2.2.1. Here, we discuss the fiber-based strain sensors that comprise CNTs. Zhang et al. developed an MWNT coated glass fiber using electrophoretic deposition (EPD) for strain sensing and electrical switching [155]. These fibers were integrated into the epoxy matrix of composites to detect microcracks that may induce catastrophic structural failure. Glass fiber substrates were used for these sensors, as glass fibers are the most widely used reinforcement materials in composites [155]. The strain sensing behavior of these fibers was divided into three regimes. At strains <1.5%, the resistance increased almost linearly due to dimensional changes in the MWNT network. Within the second region (1.5–3% strain) of strain/resistance behavior, the authors found that the resistance increased exponentially, suggesting that the space between nanotubes increased, reducing the electrical contact between the MWNTs. Finally, in the third stage of strain behavior (>3% strain), microcracks were initiated in the fiber such that MWNT networks were disconnected, and the resistance jumped to infinity. The composite was found to fracture at strains of 3.4%. These fiber sensors are minimally invasive and could be seamlessly integrated into a composite to provide real time monitoring and predict composite fracture in advance. The fabrication technique described in this work could be applied to other non-conductive fibers to serve as embedded or surface mount strain gauges for turbines, air crafts, automobiles, etc.

Several researchers have explored coating cotton or cellulosic yarns with CNTs via dip-coating. The interaction of SWNTs and cellulose yarns is largely due to van der Waals forces and hydrogen bonds. The functional groups of SWNT, such as carboxyl and hydroxyl groups, can form strong hydrogen bonds with the hydroxyl groups of cellulosic fibers [34]. Furthermore, the flexibility of SWNTs allows them to conformally adhere to the microfibrils of cellulosic fibers, thus providing maximal contact area between the SWNTs and the fibers. Kang et al. fabricated flexible, durable, and wearable sensors by coating cotton yarns with SWNTs [133]. SWNTs were purified and dispersed in a

1,2-dichlorobenzene (1,2-DCB), and cotton yarns were dipped into this solution. These SWNT-cotton yarn sensors showed a decrease in resistance with increasing strain. The negative piezoresistance of this yarn was attributed to increased mechanical fibril contacts upon strain application, which led to improved electrical pathways. The reported *GF* of −24 is greater than conventional metal strain gauges. Tai et al. also fabricated SWNT coated cotton yarn capable of strain and pressure sensing [117]. While this research was primarily focused on pressure sensing, the authors do mention the strain sensing capabilities of their fibers. The cotton fibers were coated with SWNTs via the dip-dry method, resulting in multiple layers of coatings that led to enhanced electrical conductivity. Coated fibers were then twisted to create a double-twisted smart yarn. The authors observed two regimes of strain behavior. Below 0.4% strain, SWNT-coated fibers were packed together more tightly, thus increasing fiber to fiber contact and increasing conductivity. In strain regimes greater than 0.4%, changes in resistance were attributed to the piezoresistive effect of the conductive coating. As the double-twisted yarns were strained, the distance between the nanoparticles changed, thus altering resistance. No strain sensing performance parameters are reported herein, as the focus is primarily on pressure sensing; however, the sensors are proposed for e-skins in robotic joints such as elbows, knees, and ankles. Wang et al. developed a strain sensor manufactured in a process similar to that used in commercial textile manufacturing [153]. Cotton/PU core spun yarns were coated in SWNT to create a sensory fiber. Yarn was spun with a PU core and cotton wrapper fibers. During the yarn winding process, the yarns were passed through the SWNT bath, dried, and wound onto a package in a process similar to the dyeing process in the textile industry. The sensors provided a monotonic increase in resistance with strain attributed to disruptions in the CNT network when the fiber was stretched. Resistivity of the yarns when coated >6 times was 1.68 kΩ/mm. The coated fibers could sense strains up to 300% and could be cycled nearly 300,000 times under 40% strain without noticeable breakage. Additionally, the researchers demonstrated that the fiber could operate at up to 15 Hz with 10% strain. This indicates that the material can respond quickly, which is desirable for wearable applications. While these sensors have a low *GF* of 0.65, other advantages provided by them, such as stretchability and durability in sensing, are considerable and important. Other sensor approaches, such as those employing graphene [82], may be more sensitive but lack in stretchability. The authors demonstrated that these sensors were able to detect and monitor the movement of human limbs (fingers, elbows, eye winks). Huang et al. employed a similar fabrication technique to create piezoresistive fiber sensors by coating cellulose yarns with SWNT [134]. After the yarns were formed, they were used to knit a fabric that could be strained up to 100% in any direction. Finally, the knitted fabric went through a PPy electrodeposition such that the material also acted as a supercapacitor. The behaviors of these fiber-based strain sensors are shown in Figure 12a,b. When the material was stretched, the SWNT coated cellulose yarns moved closer to each other, thus reducing the resistance of the material (strain sensing mechanism shown in Figure 12d). The resistance change of the strain sensing electrode altered the overall resistance of the textile supercapacitor, which in turn affected the discharge current. Therefore, strain changes could be monitored as a change in discharge current, as seen in Figure 12. This sensor possessed energy storage capabilities, could be self-powered, and reported excellent stability with sustained piezoresistive performance after 4400 stretch/release cycles (see Figure 12c). Additionally, it could be used as an integrated supercapacitor system for artificial intelligence and healthcare applications.

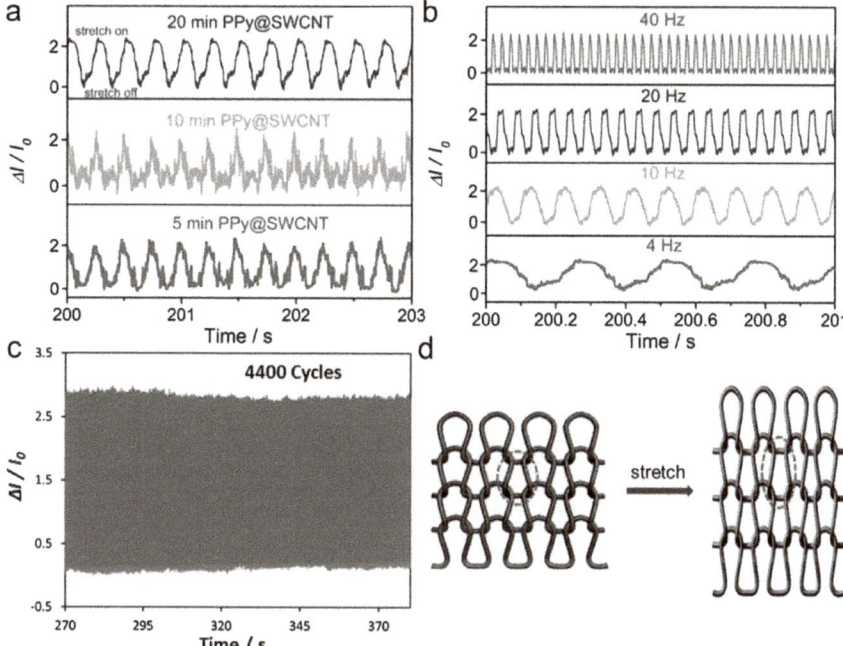

Figure 12. (a) Strain sensing performance of polypyrroles (PPy) and SWNT coated yarns in knitted configuration with different PPy deposition times. Strain may be monitored as change in supercapacitor discharge current, I, in this sensor configuration; (b) strain sensing performance of PPy and SWNT coated yarns in knitted configuration at different stretch/release frequencies; (c) durability testing of SWNT/PPy coated cotton yarn for 4400 stretch/release cycles at 40 Hz; (d) illustration of sensing mechanism for SWNT/PPy coated yarns in knitted configuration. Reprinted with permission from [134]. Copyright 2016 WILEY-VCH Verlag GmbH & Co. KGaA, Weinheim: Weinheim, Germany.

Mostafalu et al. created thread based diagnostic devices for biomedical monitoring [48]. Using microfluidic channels, several sensor types (glucose, pH, temperature, strain) were created and integrated into a thread based diagnostic device (TDD) platform. Strain sensors were created by coating carbon ink and CNTs on PU fibers. This fabrication process was discussed previously and is shown in Figure 3. The sandwich structure used to create the sensor configuration prevented CNT buckling/fracture and resulted in a linear response. When strained, the connected path of the CNTs was disrupted, leading to a rise in resistance. These sensors could measure strains up to 8% with a *GF* of ~2 and strains up to 100% with a *GF* ~3. The ability of these sensors to detect large deformations was attributed to the deformability of CNTs [28].

Pertinent details related to fiber-based strain sensors coated with CNTs are provided in Table 2. CNTs have been widely researched for strain sensing applications; however, there are challenges with achieving continuous, uniform dispersions of CNTs, and achieving proper adhesion between textile substrates and CNTs [69]. Additionally, CNTs are limited in electrical conductivity, and they tend to aggregate, which creates electrical and mechanical faults. Moreover, they are expensive. For these reasons, other carbonaceous particles have been explored for strain sensing applications and are discussed in the following sections.

Table 2. Fiber-based strain sensors coated with carbon nanotubes.

Materials	Fabrication	Sensing Mechanism	Sensor Properties	Applications
Glass fibers and MWNT [155]	EPD	Region 1: dimensional changes in MWNT network. Region 2: distance between nanoparticles increases. Region 3: microcracks in MWNT network, resistance increases.	• Conductivity ~0.1–10 S/m • Sense up to 3% strain	In-situ tracking of microcracks to predict failure
Cotton yarns and SWNT [133]	Dip coating	Negative piezoresistance due to mechanical contact between SWNT coated fibers. Increased contact, improved electrical pathways, decrease in resistance.	• Conductivity: ~80 µS/cm • GF: −24 (1–3% strain) • Linear response at <1.5% strain	• Wearable electronics. • Self-regulating resistance garment
Cotton threads and SWNT [117]	Dip coating	When strained, fiber-to-fiber contact changes, distance between nanoparticles increases, resistance changes.	• Resistance: 2×10^5–1.2×10^6 Ω/3 cm • Sense up to 2% strain	• 3D flexible pressure sensing • E-skin in soft robotic joints
PU core/cotton wrapper yarns and SWNTs [153]	Bobbin winding	When strained, disruptions in the CNT network formed, resistance increased.	• Resistance 1.68 kΩ • GF: 0.31–0.65 • Sense up to 300% strain • Durability 300,000 cycles • Fiber operated up to 15Hz with 10% strain application.	Human motion monitoring (fingers, elbows, eye winks)
Cellulose yarns and SWNTs [134]	Bobbin winding	When stretched, gaps between yarns became smaller, conductive paths were improved, resistance decreased.	• Sensed up to 100% strain • Sustained piezoresistive performance after 4400 strain cycles • Measures 100% strain at 4, 10, 20, 40 Hz	Integrated sensor/supercapacitor system for artificial intelligence and healthcare
PU threads and carbon ink/CNT [48]	Plasma treated	When strained, disruptions in the CNT network formed, resistance increased.	• GF: ~2 (8% strain), ~3% (100% strain) • Sensed up to 100% strain • Linear response to strain	Integrated with microfluidics, temperature and pH sensors as TDD

EPD: electrophoretic deposition, TDD: threat based diagnostic device.

Graphene Coated Strain Sensors

As we discussed earlier in Section 2.2.2, graphene is yet another material that has been used in coating fibers to impart conductivity. Hao et al. coated glass fibers with either MWNT or graphene and used the fibers as both temperature and strain sensors [154]. Glass fibers were coated via EPD in N_2 followed by thermal treatment to decrease resistance by ~70%. Graphene coated fibers (GF = 9.5, 2% strain) were shown to be more sensitive than those coated with MWNT (GF = 2.3, 2% strain), since electrical disconnection was more severe in the graphene coating, leading to greater sensitivity. These fibers were only able to sense in the strain range of 1.3–4%, beyond which the increase in resistance was too large to quantify. Nevertheless, these fibers can be used for detection of structural damage within composites—an application that requires low strain ranges. Cheng et al. developed a fiber-based sensor capable of sensing strain, bending, and torsion using plasma treated fibers with PU core and PET wrapper fibers [116]. Plasma treatment created polar groups on the yarn surface such that GO stuck to the yan surface during dip-coating. Finally, the yarn was soaked in HI to form rGO. When strained, the yarn resistance changed due to movement of the rGO coated PET fibers. Upon stretching, the PET wrapper fibers separated, which decreased contact area and increased resistance. These fiber-based sensors had a conductivity of 1.4×10^3 S/cm after six dip-coatings, a detection limit of 0.2% strain, a maximum strain sensing range of 100%, fast response (<100 ms), high reproducibility (up to 10,000 cycles), large rupture strain (>650%), a stable signal after 1000 cycles, as well as a GF of 10 within 1% strain and 3.7 within 50% strain. This sensor was able to monitor activities ranging from subtle motions (speech recognition and sleep quality) to rigorous activities (jogging and jumping). The results of electromechanical testing of this fiber-based sensing approach are shown in Figure 13.

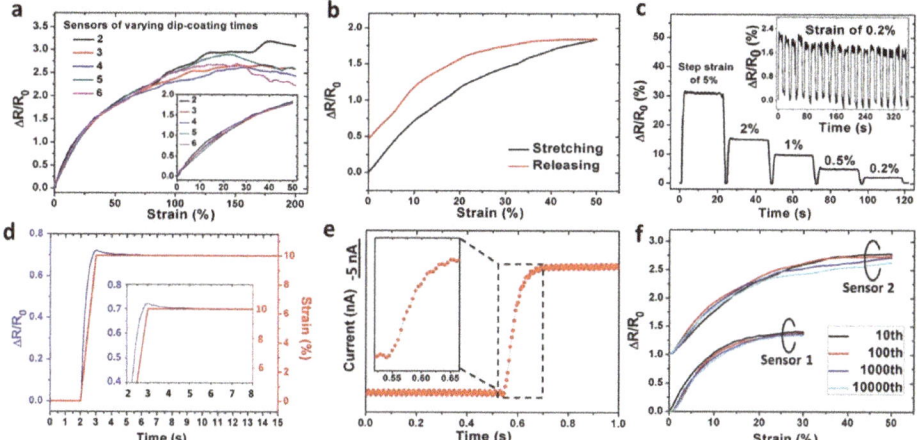

Figure 13. Sensor performance parameters of reduced graphene oxide (rGO) coated fibers: (a) Influence of coating number on relative resistance change when strained 200%; (b) resistive response to strains under 50%; (c) resistance/strain behavior under various levels of strain ranging from 0.2–5% strain; inset shows sensor behavior when 20 cycles of 0.2% strain were applied; (d) variation of resistance upon 10% strain application; inset shows a closeup of the overshoot; (e) fiber sensor signal in response to quasi-transient input strain of 0.5%; inset shows a close up of the response time; (f) results of 10,000 cyclic testing with 10th cycle (black), 100th (red), 1000th (blue), and 10,000th (cyan) cycles at 0–30% (sensor 1) and 0–50% (sensor 2). Reprinted with permission from [116]. Copyright 2015 WILEY-VCH Verlag GmbH & Co. KGaA, Weinheim: Weinheim, Germany.

Park et al. created a highly stretchable and wearable graphene fiber sensor [82]. The authors applied several dip coatings in an lbl method to commercially available yarns in a dip coating process. Three yarn types—RYs, NCRYs, and WYs—were dipped in GNPs. This fabrication method was discussed previously and is shown in Figure 8. The sensing behavior varied depending on the yarn type. RYs and NCRYs exhibited an increase in resistance in strain, while WYs resistance decreased with strain. RYs formed cracks when stretched, which broke electrical paths and greatly increased resistance. NCRYs were made such that a rubber core was wrapped by nylon fibers. GNPs were adsorbed by the outer nylon fibers, while the inner fibers were only partially coated. Under strain, nylon fibers were not stretched but instead deformed in the direction of strain, thus there was less cracking, resulting in a reduced rate of resistance change. This was reflected in the GFs with the RYs displaying an exponential increase in resistance, while the NCRYs provided greater range of sensing but a lower sensitivity. WYs comprised crimped and twisted fibers that straightened during strain application, which increased fiber contact. This created increased electrical paths such that resistance decreased with strain. RYs and NCRYs required PDMS coating to avoid GNP delamination. These yarns could be stretched up to 150% and showed stability and fast response when tested with 10 cyclic tests of strains up to 80% (rubber), 100% (nylon/rubber), and 40% (wool). These fibers were shown to be capable of monitoring both large and small human motions.

Zhang et al. developed a strain based sensor with a Meyer rod coating technique [85]. Spandex, PP, silk fibers, and human hair were coated with dry graphite flakes using a Meyer rod. Ten conductive coatings were applied, and fibers were encapsulated in silicone. Sensing capabilities were explained by the overlapping area between graphite flakes. When stretched, the overlap area decreased, leading to a rise in resistance. Once the fiber was allowed to relax, the graphite contacts returned to their initial states, providing reproducible electrical behavior. These fiber sensors provided varying *GF*s depending on fiber type. *GF*s of 71.1 (up to ~10% strain), 14.2 (up to ~18% strain), 14.5 (up to ~15% strain), and 14.0 (up to ~30% strain) were reported for human hair, silk, PP, and spandex fibers, respectively. Silk fibers—the primary focus of this work—provided low drift, low hysteresis, stability during >3000 stretching-releasing cycles at 10% strain, and a response time of 135 ms. Interestingly, the sensor displayed electrical responses only dependent on strain level instead of frequency (which was tested at frequencies ranging from 0.25–6 Hz), implying that the rate at which strain is applied does not affect the output signal, indicating high reliability. These sensors were shown to be capable of monitoring joint motion. Multiple sensors were connected to the human body in different directions to measure strains along multiple axes to monitor intricate body deformations.

Pertinent details related to fiber-based strain sensors coated with graphene are provided in Table 3. Graphene based strain sensors are advantageous due to their high conductivity, high gauge factors, and low cost [77]. However, many graphene based sensors are not very stretchable [153]. Additionally, research is still needed to produce rGO with uniform properties, especially for large-scale use in large-area fiber sensors. Furthermore, many graphene-based strain sensors require encapsulation to avoid delamination due to the weak interface between fiber and graphene [82]. Encapsulation layers increase the bulk of fiber-based sensors and take away from the textile-like feel of fiber-based sensors, which is desirable in wearable technology.

Table 3. Fiber-based strain sensors coated with graphene.

Materials	Fabrication	Sensing Mechanism	Sensor Properties	Applications
Glass fibers, CNTs or graphene [154]	EPD	When strained, broken conductive pathways resulted in increased resistance. Disconnection of graphene was more severe, created larger resistance change compared to CNT coated fibers.	• Resistance: 7.4–25.4 MΩ • GF: 2.3 (CNT, strain 2%), 9.5 (graphene, strain 2%) • Sensed up to 2% strain	Embedded composite epoxy to detect damage in structural composites
PU core/PET wrapper fiber yarns and graphene oxide (GO) [116]	Dip-coating	When strained, GO/PET winding fibers separated, contact area decreased, resistance increased.	• Conductivity: ~0.15 S/m • GF: 10 (1% strain), 3.7 (50% strain) • Response unchanged after 1000 cycles • Detection limit 0.2% strain • Response time: <100 ms • Reproducible signal up to 10,000 cycles	• Health and motion monitoring • Detecting full-range human activities
Rubber, nylon, wool yarns and GNPs [82]	Dip-coating	GNP coating on RY & NCRY cracks when strained resulting in broken electrical pathways, and increased resistance. When WY is strained there is greater contact between coated fibers, increasing conductivity.	• GF: ~1–2.5 (NCRY, strains <150%), ~6–15 (RY, strains <50%), <~1 (WY, strains <50%) >Sensed up to 150% (NCRY), 25% (RY), 50% (WY)	Human monitoring–small and large motions
Spandex, PP silk, human hair and graphite flakes [85]	Meyer rod.	When strained, overlap area between graphite flakes decreased, resistance increased.	• GF: 71.1 (~10% strain), 14.2 (~18% strain), 14.5 (~15% strain), 14.0 (~30% strain) for human hair, silk, PP, spandex respectively • Silk: low drift, hysteresis >3000 cycles • Silk Response time: 135 ms • Frequency independent response from 0.25–6 Hz	• Monitoring joint motion. • Multiple sensors able to measure complicated body movements

Other Carbon Coated Strain Sensors

Other carbonaceous coatings, excluding CNTs and graphene, were previously discussed in Section 2.2.3. Here, fiber-based strain sensors employing CB are discussed. Ge et al. developed a stretchable electronic fabric-based artificial skin [126]. This work was based on work by previous research [157] in which a large, stretchable, silicon nanoribbon network was used to create an artificial skin. While this approach was unique and provided desirable performance, the use of silicon is expensive and prevents wide acceptance. Therefore, Ge et al. developed a low-cost alternative. Commercial elastic threads helically wrapped in nylon fibers were employed. Yarns were coated in silver nanowire (AgNWs) dispersion through dip-coating. The AgNW layer of this fiber acted as an electrode and was coated in a piezoresistive rubber (PDMS loaded with CB) via dip-coating. The piezoresistive layer was responsible for the sensory characteristics of this fiber. When a mechanical stimulus was applied to the fibers, a change in contact area, thickness, or conductivity of the rubber layer was induced, which altered the electrical resistance. In the case of strain, the thickness of the conductive layer changed, thus allowing strain to be monitored as a change in resistance. The pressure sensing capabilities of these fibers was characterized extensively, while strain sensing was discussed minimally. The potential for these sensors to act as large-area artificial skins capable of monitoring and differentiating pressure, strain, and flexion was demonstrated.

Wu et al. created strain sensors by a sequential lbl coating process in which PU yarns were dipped in a conductive polymer composite containing CB and natural rubber [156]. EPD was used to combine CB with cellulose nanocrystals to stabilize CB and create a negatively charged material. Subsequently, the CB composite was combined with natural rubber and remained negatively charged. PU yarns were dipped into a chitosan solution, which created positive charges on the yarn surface. The positively charged yarns were dipped into the negatively charged CB composite, thus coating the PU yarn surface. These sensors provided a large *GF* (39.1), low detection limit (0.1% strain), good reproducibility (>10,000 stretch/release cycles at 1% strain), and excellent wash and corrosion resistance. However, the yarns were not tested beyond 10% strain. Additionally, the conductivity of the fibers was low at 4.1 MΩ/cm with 50 coating layers. However, this could be improved by increasing the CB percentage. The authors demonstrated that the fibers were able to differentiate between speech, monitor pulse, detect emotion (via monitoring of facial motion), and monitor finger motion. Thus, the fibers were proposed for artificial intelligence products. Pertinent details related to fiber-based strain sensors coated with carbonaceous materials are provided in Table 4. Carbonaceous coatings such as CB are advantageous due to their cost efficiency compared to materials such as graphene, SWNT [156], and silicon [126], as well as their relatively simple coating techniques. However, carbonaceous coatings do not provide excellent electrical conductivity, low contact resistance, or structural stability [24]. Metal-based coatings, discussed in the following section, have been explored to overcome these issues.

Table 4. Fiber-based strain sensors coated with other carbonaceous materials.

Materials	Fabrication	Sensing Mechanism	Sensor Properties	Applications
PU core/nylon wrapper fiber yarns coated in silver nanowires (AgNWs), PDMS/CB rubber [126]	Dip-coating	When strained, the piezoresistive rubber layer changed thickness, resistance increased.	• Conductivity: 1600 S/cm • Sensed up to 100% strain • Capable of sensing pressure, lateral strain, and flexion	• Health monitoring • Artificial skin for robotics • Biomedical prostheses • Physiological analysis device
PU yarns and polymer/CB composite [156]	Dip-coating/EPD	When strained, CB/composite layer formed cracks, resistance increased.	• Yarn resistance: 4.1 MΩ/cm • GF: 39 (1% strain) • Sensed up to 3% strain • Strain Detection limit: 0.1% • Reproducibility—over 10,000 cycles (1% strain) • Wash and corrosion resistance	• Human motion monitorin • Monitoring minute physiological activity

PDMS: polydimethylsiloxane.

3.3.3. Metal Coated Strain Sensors

Metallic coating techniques were previously discussed in Section 2.3. Here, fiber based strain sensors utilizing metal coatings are discussed. Wei et al. developed highly conductive, stretchable, and sensitive sensors using PU fibers coated with AgNWs using a Chinese brush pen [114]. AgNWs were dispersed in a waterborne PU (WPU) to create the silver ink. Intermolecular hydrogen bonding between the AgNWs and the WPU allowed them to immobilize more securely on the PU fiber such that AgNWs did slip excessively during straining. Following coating, the prestrained fibers were allowed to relax, and a unique microstructure on the surface of the fiber was formed. The surface texture of the ink helped explain many of the unique fiber properties. As the fiber was strained, three regions of resistive sensing behavior were observed. In the lowest strain region, the resistance was well sustained, and the wrinkled microstructure gradually flattened. In the second strain region, AgNWs deformed, and the conductive network resistance gradually increased. Finally, in the third strain/resistance regime, excessive slippage of the AgNWs ruptured the conductive network such that resistance increased rapidly. These fiber sensors were well characterized with high conductivity (10^{-4}–10^{-5} Ω cm), high stretchability (400%), high durability (more than 1200 s ultrasonic treatment), and showed an ability to withstand bending deformation of 0.12 rad^{-1} with response and relaxation times of 35 ms and 15 ms, respectively, as well as excellent working stability (>4000 loading/unloading cycles). Many of the sensing capabilities of this fiber were attributed to the surface microstructure.

Similarly, Chen et al. proposed a PU/AgNW strain sensor for electronic data gloves, as shown in Figure 14g,h [115]. Commercially available yarns made up of a rubber latex core and PU winding fibers were coated with poly(vinylidenefluoride-co-trifluoroethylene) (P(VDF-TrFE)) polymer nanofiber mat through electrospinning. Then, the yarns were dip-coated in a AgNW dispersion. The P(VDF-TrFE) layer was employed as a protective and enhancing layer (enhancing stability of electrical conductivity), while the AgNW layer was responsible for conductivity. Throughout coating processes, the yarns were prestrained to optimize coverage and to form a wrinkled structure upon release. The wrinkled structure protected the conductive layer and prevented breakage under strain. The sensor provided a GF of 5.326 (within 25% strain), rapid response time of 20 ms, durability after 10,000 cycles (90% of sensitivity was maintained indicating low drift), and conductivity of 5 Ω/cm (seven dip coatings). A slow and monotonic increase in resistance was observed up to 60% strain, and then a rapid increase was observed after that. This was attributed to the fact that, at lower strains (<60%), the wrinkles of the conductive layer spread (wrinkles removed), while at larger strains, the conductive layer actually deforms, causing a more rapid change in resistance. These fiber sensors could detect bending (see Figure 14) and torsion and were proposed for human–machine interfaces. In related work, Cheng et al. created a fiber-based sensor for e-skin applications with AgNW coating on PET-wrapped PU yarns [118]. When strained, the AgNW network was cracked, which in turn increased electrical resistance. The cracking mechanism of the AgNWs could be related to not only the amplitude of strain but also the direction. These sensors were able to detect strains up to 40–50%, beyond which the fiber was permanently damaged. In strain ranges less than 30%, the fiber provided a *GF* of 3.2. Pertinent details related to fiber-based strain sensors coated with metallic particles are provided in Table 5. To the authors knowledge, all fiber-based strain sensors made via metal coating processes utilize AgNWs. AgNWs, which are easily synthesized, flexible, and highly conductive, may be promising candidates for flexible electronics [114,158].

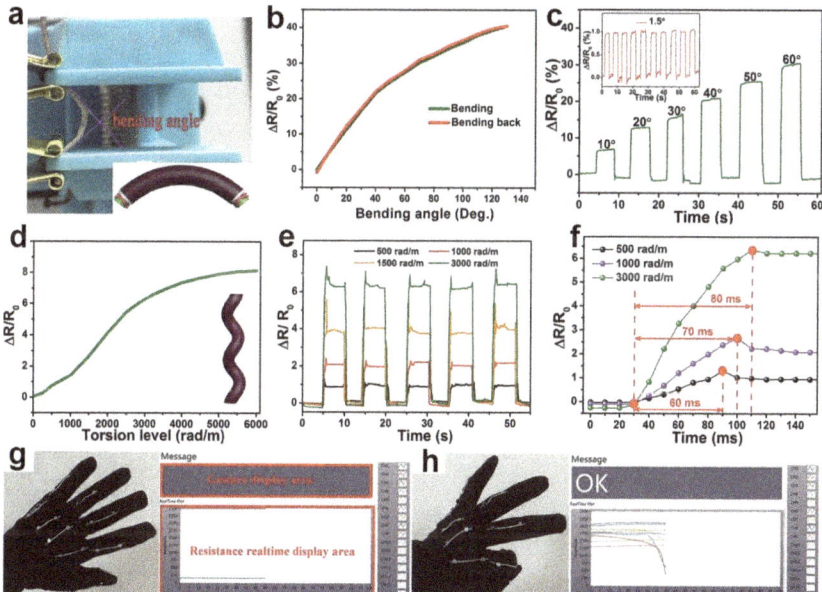

Figure 14. Bending and torsion sensing behavior of PU fiber coated in poly(vinylidenefluoride-co-trifluoroethylene) (P(VDF-TrFE)) and AgNWs: (**a**) Fiber sensor bending test setup and bending angle defined; (**b**) resistance response when fiber bent in forward and backward directions; (**c**) change in fiber resistance upon application of various bend angles. Inset: cyclic testing to 1.5° bending angle; (**d**) change in fiber resistance upon application of torsion test up to 6000 rad m^{-1}; (**e**) results of cyclic torsion testing under varying degrees of torsion; (**f**) response time testing for torsion detection; (**g**) "data glove" produced with strain sensing fibers and corresponding software to indicate finger movement; (**h**) "data glove" providing different output corresponding to hand gesture. Reprinted with permission from [115]. Copyright 2016, John Wiley and Sons: Hoboken, NJ, USA.

Table 5. Fiber-based strain sensors coated with metals.

Materials	Fabrication	Sensing Mechanism	Sensor Properties	Applications
PU yarn and AgNW ink [114]	Chinese brush pen	Region 1: resistance was well sustained before the wrinkled microstructures flattened. Region 2: AgNWsslippage deformed the conductive networks and increased resistance. Region 3: excessive AgNWs slippage ruptured conductive networks and increased resistance.	• Conductivity: 10^{-4}–10^{-5} Ω/cm • Sensed up to 400% strain • Durability: more than 1200 s ultrasonic treatment • Withstood bending deformation of 0.012 rad^{-1}	• Wearable devices • Robotic systems • Smart fabrics
Rubber core/PU wrapper yarns, P(VDF-TrFE) nanofiber mat, AgNWs [115]	Dip-coating	Region 1: resistance increased monotonically and slowly, wrinkled microstructures became gradually flattened. Region 2: slippage of AgNWs deformed the conductive networks, resistance gradually increased.	• Resistance: 5 Ω/cm • GF: 5.326 (25% strain) • Sensed up to 60% strain • Durability: up to 10,000 strain cycles • Response time: 20 ms • Can detect bending angles up to 130°	• Monitoring finger motions and subtle physiological signals • Smart glove for real-time gesture recognition
PU core/PET wrapper fiber yarns and AgNWs [118]	Dip-coating	When strained, AgNW layer cracked, resistance increased.	• GF: ~3.2 (30% strain) • Sensed up to 50% strain • Strain detection limit 1%	• E-skin

3.4. Coatings for Fiber-Based Pressure Sensors

Fiber-based pressure sensors made via coatings methods are not as prominent as those used for strain sensing; however, some efforts have been made in this field [24,26,136]. While coating of entire fabrics to create pressure sensors has been explored [159–165], this discussion is limited to fiber and yarn based sensors. Pressure based fiber sensors generally work in capacitive modes [24,26,118,119,127], while few studies have explored resistive [114,117] sensing for pressure detection. Fiber-based pressure sensors have been proposed for applications such as electronic skins [126], wearable health sensors [136], and robotics [119].

3.4.1. Intrinsically Conducting Polymer Coated Pressure Sensors

ICPs are attractive alternatives to carbonaceous materials because of their high conductivity and facile processing properties of polymers. Takamatsu et al. created a relatively large (16 cm × 16 cm) fabric sensor using PEDOT:PSS coated nylon fibers for detecting pressures [136]. Nylon fibers were unwound from a package and passed through a nozzle that coated the fiber surface with PEDOT:PSS. These nozzles were designed such that the coating thickness was easily controlled, followed by drying and collection on a second package winder. This method was proposed to create long lengths of coated fibers. Alternative coating methods that require vacuum chambers for coatings cannot provide adequate fiber lengths for commercial textile processes. After fiber fabrication, the fiber-sensors were woven into an array (Figure 15b,c), and the ability of the array to map pressure was demonstrated (see Figure 15d). In a woven configuration, the PEDOT:PSS layers acted as electrodes separated by fluoropolymer (Cytop) dielectric layers. This design allowed the fibers to act as electrodes in a capacitive configuration. When pressure was applied, the overlapping electrode area increased, thereby increasing the capacitance, see Equation (2). The fibers provided a sensing range of 0.98–9.8 N/cm^2, which allowed for sensing human touch. This sensing approach is suggested for healthcare and wearable keyboard applications. This is the only publication to our knowledge that explored coating fibers for pressure sensors, and its pertinent sensor properties are listed in Table 6. The following section discusses the use of carbonaceous particles for pressure sensors made via coating.

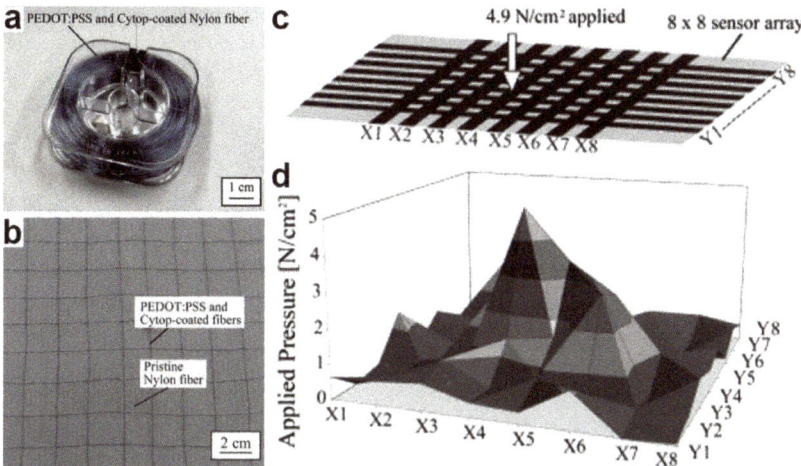

Figure 15. (a) 100 m long PEDOT:PSS coated nylon fiber; (b) 16 cm × 16 cm fabric array produced with PEDOT:PSS coated nylon fibers; (c) demonstration of a sensing 8 × 8 pixel array; (d) detected pressure on textile based sensing array. Reprinted with permission from [136]. Copyright 2012 Elsevier B.V.: Amsterdam, The Netherlands.

Table 6. Fiber-based pressure sensors coated with intrinsically conducting polymers and carbonaceous materials.

Materials	Fabrication	Sensing Mechanism	Sensor Properties	Applications
Nylon fiber, PEDOT:PSS Cytop (dielectric) [136]	Nozzle/die-coating	Fibers woven into array. When pressure applied, overlapping electrode area was increased, thus causing a change in capacitance.	• S: 0.98–9.8 N/cm^2 • Lengths adequate for commercial weaving	• Healthcare • Wearable keyboards
Cotton threads coated and SWNT [117]	Dip-coating	When loaded, fiber-to-fiber contact changes, distance between nanoparticles increases, resistance changes.	• S: 0.1–1.56% kPa^{-1} • Stability: >10^4 cycles • Response time of 2.5 Hz • Capable of measuring cyclic load of 25 kPa at 1 Hz • Capable of predicting both the intensity and the position of an applied load	• 3D flexible sensor • E-skin in robotic joints
PU core/nylon wrapper fiber yarns coated in AgNWs, PDMS/CB rubber [126]	Dip-coating	When strained, the piezoresistive rubber layer changed thickness, resistance increased.	• Conductivity: 0–1600 S/cm • S: 4.29 N^{-1} (~0–0.4 N), 0.02 N^{-1} (~0.4–2 N) • Stability: 100m000 load /release cycles (0.03 N) • Stretchability ~100% strain • 8 ms delay time at 60 Hz, 5 ms for 5 Hz pressure • Capable of measuring load when stretched	• Health monitoring and disease diagnostics • Artificial skin for robotics • Biomedical prostheses

3.4.2. Carbon Coated Pressure Sensors

Carbonaceous structures and coating techniques were previously discussed in Section 2.2. Here, fiber-based pressure sensors coated with carbonaceous particles are discussed. A previously discussed fiber strain sensing approach [117] was also explored for pressure sensing applications. Cotton fibers were coated with SWNT through a dip-coating method. Coated fibers were twisted to create a double-twisted smart yarn. Pressure sensing capabilities were attributed to the fact that, when yarns were loaded, the fiber-to-fiber contact increased, while the distance between nanoparticles also increased, thus altering electrical resistance. These twisted yarn structures provided sensitivity values ranging from 0.1–1.56%, a response time of 2.5 Hz, the capability of measuring pressures up to 25 kPa at 1Hz, good resilience and fatigue properties when exposed to >10,000 loading cycles, as well as the ability to predict both the intensity and the position of applied load. The yarns were proposed for applications including e-skins and robotic joints (elbows, knees, and ankles). Similarly, the fiber-based strain sensors previously discussed by researchers Ge et al. also operate as pressure sensors [126]. Commercial elastic filaments with inner PU core and helically wrapped nylon fiber sheath were dip-coated with AgNWs, followed by coating with a piezoresistive PDMS-CB composite rubber. The piezoresistive rubber layer was responsible for the sensing capabilities of these fibers. Fibers were then woven into an array for pressure sensing. When a mechanical stimulus was applied to the fibers, a change in contact area, thickness, or conductivity of the rubber layer was induced, which altered the electrical resistance. In the case of pressure sensing, the authors noted two regimes of sensing behavior. In the first pressure regime (0 to ~0.4 N), the authors attributed changes in resistance to changes in fiber contact area with pressure application. When larger loads were applied to the fiber array (0.4–2 N), the conductivity of the piezoresistive rubber layer increased due to CB particles moving closer with load application. These fiber sensors provided a conductivity in the range of 0–1600 S/cm, stretchability up to 100% strain, sensitivity of 4.29 N^{-1} (region 1: 0–~0.4 N) and 0.02 N^{-1} (region 2: ~0.4–2 N), stability/durability sufficient to withstand 100,000 load/release cycles (0.03 N), a response speed of 8 ms during pressure sensing, load measuring capabilities in a stretched state, as well as the ability to measure strain, bending, and twisting. The results of electromechanical tests of these fibers are shown in Figure 9e–j. These fibers are suggested for large-area artificial skin capable of monitoring and differentiating pressure, strain, and flexion. Pertinent details related to fiber-based pressure sensors coated with carbonaceous materials are provided in Table 6. Carbon based coatings have not been explored extensively for pressure sensors. A more active area of research is the use of metallic coatings for pressure sensors, which is discussed in the following section.

3.4.3. Metal Coated Pressure Sensors

Metallic coating techniques were previously discussed in Section 2.3. Here, fiber-based pressure sensors utilizing metal coatings are discussed. An early work in this field was by Hasegawa et al. in which hollow rubber fibers were coated with gold films to produce capacitive sensors when fibers were placed in a woven configuration [127]. The fiber surface was sputtered by argon and oxygen plasma to improve adhesion. This was followed by a sputtered deposition of gold film with the fibers rotating in the vacuum chamber to ensure complete surface coverage. Finally, the fibers were coated with an insulating layer of parylene-C. Upon pressure application, fiber crossover points were compressed such that the contact area was increased, resulting in an increase in capacitance. A single fiber made in this way could elongate up to 10% and provided a stable response (less than 5% hysteresis) when exposed to three cycles of 300 mN force. The fibers were incorporated into gloves to demonstrate their pressure sensing capabilities. Lee et al. developed a cotton-based sensor coated with platinum using low temperature ALD and PDMS [26]). Similar to previously discussed research, when fiber cross-over points were formed via sewing, the platinum (Pt) layers served as electrodes, and PDMS acted as the separating dielectric. When pressure was applied to the fibers, the dielectric layer thickness decreased and contact area increased, resulting in a change of capacitance. These fibers had a resistance of 2.72 Ω/cm, were able to detect loads up to 30 mg, and provided stable response when exposed to 100,000 cycles of 0.4 N load.

A previously discussed fiber strain sensing approach [114] was also explored for pressure sensing. These fibers were subsequently twisted into a structure capable of detecting pressure and bending deformations. When pressure was applied to the twisted fiber structure, the contact area between the fibers increased, therefore causing electrical resistance to increase. These fiber sensors were well characterized with high conductivity (10^{-4}–10^{-5} Ω cm), a sensitivity of 0.12 kPa^{-1} (see Figure 16g), high stretchability (400%), high durability (more than 1200 s ultrasonic treatment), an ability to withstand bending deformation of 0.012 rad^{-1}, a response time of 35 ms, a relax time of 15 ms, and excellent working stability (>4000 loading/unloading cycles). Many of the sensing capabilities of this fiber were attributed to the surface microstructure. Similarly, the strain sensor previously discussed by Cheng et al. was also demonstrated as a pressure sensor [118]. The fibers were arranged into an array and integrated into a PDMS film. When pressure was applied, the thickness of the dielectric PDMS layer decreased, thus altering capacitance. These fiber-based capacitive pressure sensors provided a sensitivity of 0.096 KPa^{-1} (<0.1 kPa) and 1.1 MPa^{-1} (0.1–10 kPa), a dynamic sensing range of 0–50 kPa, a detection limit of 1.5 Pa, a response time of 32 ms, and an ability to withstand 10,000 cycles of 0.84 and 13.7 kPa pressure application with little drift (3.8–4.6%). However, the conductive coating on this fiber was damaged if stretched above 50%. Corresponding electromechanical test results are shown in Figure 17.

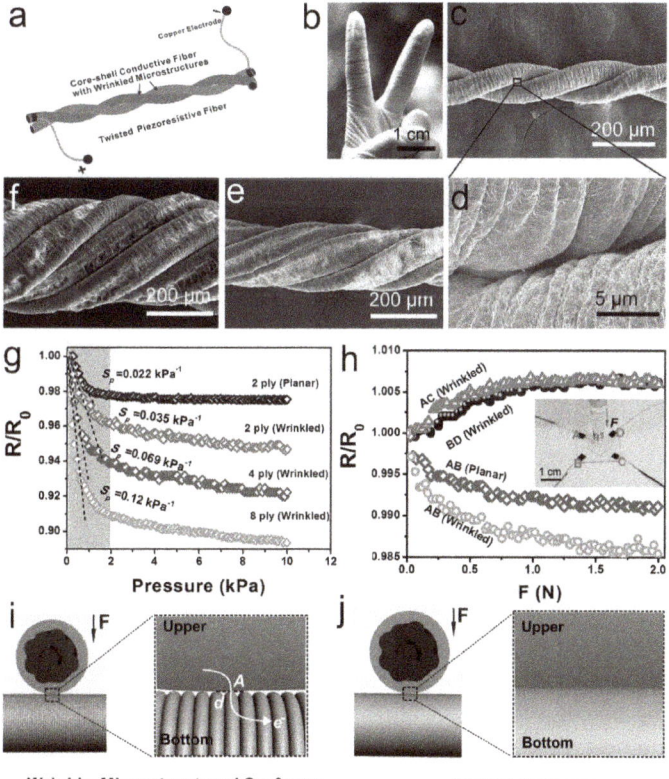

Figure 16. (a) AgNW coated PU twisted fiber structure; (b) demonstration of sensory fiber flexibility and elasticity as fiber is wrapped around finger several times; (c) SEM of two AgNW coated PU fiber twisted into yarn configuration with (d) showing the SEM of selected area of (c); (e) SEM of four AgNW coated PU fiber twisted into yarn configuration; (f) SEM of eight AgNW coated PU fiber twisted into yarn configuration; (g) strain/resistance performance of yarns that comprise different numbers of coated fibers; (h) mechanical testing of samples with wrinkled microstructure and non-wrinkled microstructure (planar) to investigate sensing mechanism; (i,j) schematic illustrations of sensing mechanisms where "A" denotes contact area and "d" indicates distance between conductive coatings. Reprinted with permission from [114]. Copyright 2016 WILEY-VCH Verlag GmbH & Co. KGaA, Weinheim: Weinheim, Germany.

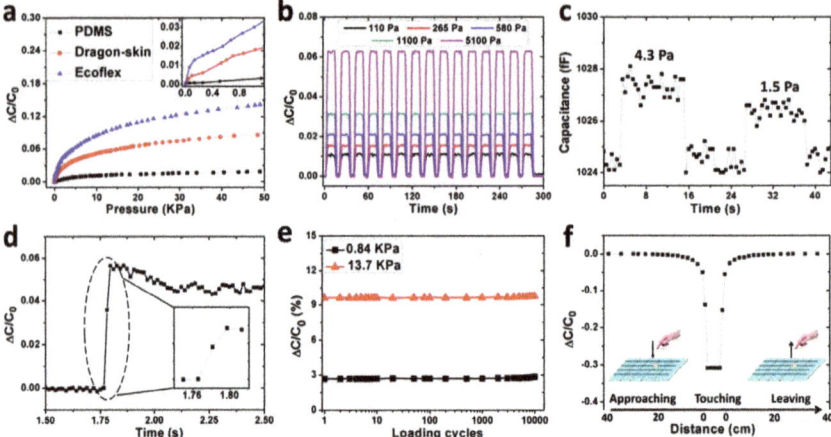

Figure 17. (a) Pressure versus capacitance of fiber sensors embedded in various PDMS dielectric layers. Inset shows response within low pressure regions of 1 kPa; (b) capacitive response of fibers when exposed to pressures ranging from 110–5100 kPa; (c) application of pressure steps of 4.5 Pa and 1.5 Pa to determine the detection limit; (d) real-time capacitance response to 1.5 kPa pressure application. Inset shows a close-up of the response; (e) durability testing of 10,000 cycles of 0.84 and 13.7 kPa; (f) relative change in capacitive response when human finger approaches array, touches array, and leaves array. Reprinted with permission from [118]. Copyright 2013 Royal Society of Chemistry: London, UK.

Lee et al. produced a fiber based sensor with Kevlar fibers [24]. The fabrication process for these fibers is described in Section 2.3. The silver nanoparticle (AgNP) layers acted as the electrodes of the capacitor, while the PDMS layers formulated the dielectric. When pressure was applied, increased contact area and reduced dielectric thickness increased capacitance. These fiber-based pressure sensors had a conductivity of 0.15 Ω/cm, a sensitivity of ~0.21 kPa^{-1} (<2 kPa), ~0.064 kPa^{-1} (>2kPa) stability during 3000 bending cycles (180°) and 10,000 press/release cycles (1.7 N), a response time of ~40 ms, a relaxation time of ~10 ms, a maximum detection limit of 3.9 MPa, as well as negligible hysteresis. The fibers were incorporated into a fabric to show their ability to act as textile-based sensors and were used as a pressure sensitive controller for a drone. Similar research was completed by researchers Chhetry et al. [119]. First, Twaron fibers were coated with SBS then immersed in an AgNP precursor and subsequently reduced using an $N_2H_4 \cdot 4H_2O$ solution. AgNP coated fibers were then coated with a PDMS dielectric material such that, when two fibers were stacked perpendicularly, a capacitive sensor structure was formed. A large focus of this work was on altering the microstructure of the dielectric PDMS to optimize sensing behavior. Glucose particles were added to the PDMS and then removed after curing to create pores within PDMS, providing enhanced effective permittivity and lower stiffness. Various PDMS/glucose ratios were studied to determine the effect of dielectric porosity on sensing capabilities. When pressure was applied to the fibers in a stacked configuration, the dielectric layer was compressed such that the AgNP electrode layers were closer and contact area increased. Additionally, the effective permittivity of the porous dielectric layer changed with pressure application due to the removal of air from the porous structure. Changes in the dielectric thickness, the electrode area, and the dielectric permittivity induced changes in capacitance, which could then be used to quantify the amount of pressure applied. These fiber sensors had a resistance of 0.1431 Ω/cm, sensitivity values of 0.278 kPa^{-1} (<2 kPa), 0.104 kPa^{-1} (2–10 kPa), and 0.0186 kPa^{-1} (>15 kPa), a dynamic range of 0–50 kPa, hysteresis of 6.3%, a response time of ~340 ms, a detection limit of 4 mg or 38.82 Pa, and cyclic testing of 10,000 cycles of 38.81 kPa and 17.43 kPa in which little drift was observed. These capacitive fiber-based pressure sensors are proposed for applications including touch panels, human–machine interfaces, health monitoring, and robot components. Pertinent details related to fiber-based pressure sensors coated with metallic particles are provided in Table 7. Metal coatings have been thoroughly explored for fiber-based pressure sensors when compared to ICPs and carbonaceous materials. Interestingly, a variety of metallic materials have been employed for pressure sensing, including gold, Pt, AgNWs, and silver nanoparticles (AgNPs). These metallic based pressure sensors do provide high conductivity and what is reported as adequate sensitivity levels [114].

Table 7. Fiber-based pressure sensors coated with metals.

Materials	Fabrication	Sensing Mechanism	Sensor Properties	Applications
Hollow rubber fiber, gold (electrodes) [127]	Sputtering	When load applied to fiber crossover point, contact area between the warp and weft fibers increased, resulting in change in capacitance.	• Max elongation of 10% (20 mN) • Stability: 3 cycles, negligible hysteresis (<5% at ~300 mN)	Tactile sensors in bedding, clothes, chairs, and shoes
Cotton fibers, Pt (electrodes) PDMS (dielectric) [26]	Low temp ALD	When load applied to fiber crossover point, contact area between the fibers increased. Additionally, dielectric thickness decreased, resulting in change in capacitance.	• Resistance 2.72 Ω/cm • Stability: Minimal drift 100,000 cycles • Able to detect 30 mg	Textile electronics
PU yarn and AgNW ink [114]	Chinese brush pen	Two fibers twisted together for pressure sensing. When pressure applied, contact area between twisted fibers increased, resistance decreased.	• S: 0.12 kPa^{-1} • Withsood bending up to 0.012 rad^{-1} • Response time: 35 ms • Detection limit 10 mg • Stability >4000 loading/unloading cycles	• Wearable devices • Robotic systems • Smart fabrics
PU core/PET wrapper fiber yarns and AgNWs, silicone (dielectric) [118]	Dip-coating	When load applied to fiber crossover point, dielectric thickness decreased, resulting in change in capacitance.	• S: 0.096 KPa^{-1} (<0.1 kpa) and 1.1 MPa^{-1} (0.1–10 kpa) • Stability: 10,000 cycles pressure application • Stretchability: up to 30–40% • Dynamic sensing range up to 50 kPa • Detection limit of 1.5 Pa • Response time ~32 ms	E-skins
Kevlar fibers, SBS, AgNPs (electrodes), PDMS (dielectric) [24]	Coating	When load applied to fiber crossover point, contact area between the fibers increased. Additionally, dielectric thickness decreased, increasing capacitance.	• Resistance: 0.15 Ω/cm • Stability 3000 bending cycles (180°), 10,000 cycles (1.7 N) • S: ~0.21 kPa^{-1} (<2 kPa), ~0.064 kPa^{-1} (>2 kPa) • Response time: ~40 ms • Relaxation time: 10 ms • Maximum detection limit of 3.9 Mpa	Human-machine interface
Twaron fibers, SBS, AgNPs, PDMS (dielectric) [119]	Dip-coating	When fiber crossover point is loaded, contact area between the fibers increased. Additionally, dielectric permittivity changes due to pores within the dielectric collapsing, resulting in change in capacitance.	• Resistance: 0.14 Ω/cm • S: 0.278 kPa^{-1} (<2 kPa), 0.104 kPa^{-1} (2–10 kPa), 0.0186 kPa^{-1} (>15 kPa) • Dynamic range: 0–50 kP • Negligible hysteresis 6.3%, • Response time ~340 ms • Detection limit: 4 mg/38.82 Pa • Stability: 10,000 cycles	• Touch panels • Health monitoring • Artificial robot arms.

ALD: atomic layer deposition, SBS: poly(styrene-block-butadiene-styrene), AgNPs: silver nanoparticles.

4. Conclusions

This review demonstrates the versatility of materials and processes for conductive coatings for textiles. It highlights the numerous applications in strain and pressure sensing regimes for electrically conductive fibers and yarns. While there is a variety of coating types and application techniques that can be used to make such fiber and yarn based sensors, it is important that this is done without sacrificing the inherent qualities that textiles possess. Textiles are soft, comfortable, conformable, washable, and lightweight. Any coating technique or method that is used must impart electrical conductivity and good sensing performance without sacrificing these qualities. Keeping that in mind, it can be concluded that some materials, e.g., conducting polymer composites composed of carbonaceous particles such as CNTs and graphene, might be more suitable than ICPs and metals, which are inherently brittle as coatings. Moreover, certain coating techniques such as dip-coating may be preferred over CVD and ALD techniques, since the latter can prove to be difficult to execute on specific fibrous materials. Hence, the conditions that the fiber substrate can withstand and the performance requirements of the sensor are also important factors when selecting the coating material and method. It is also important to keep in mind that large-scale and large-area integration of sensing into fibers requires a coating method that can be used, preferably, in a continuous roll-to-roll operation to produce a uniform coating on large lengths of yarns and fibers.

Capacitive and resistive sensors are the two most popular sensing modalities that can be integrated into the fiber structure via coating. This is due to the simplicity of the principle as well as the inherent shape of the fiber itself; akin to a strain gauge, it can act as a one-dimensional (1D) wire that can change its resistance when strained or pressed. For capacitive sensors, parallel plate configurations can be achieved by integrating coated fibers into a woven fabric configuration, whereas cylindrical capacitors can be integrated into the individual fiber or yarn structure itself. Hence, the serendipitous combination of hierarchical textile structures with electrically conductive coatings offers the possibility of integrating sensors in a conformal configuration to monitor biometrics, offer new possibilities of communication, and allow the development of new classes of soft robots.

In summary, the ideal material and application specific coating material should offer a balance in terms of processability, sensing performance, environmental stability, etc. While there are certain limitations to coating fibers—such as delamination of the coating material, cracks developing within the coating layer, lack of uniformity in coating thickness, and the migration of the coating material into the fiber itself—by judicious choice of appropriate coating methods and materials, such limitations can be overcome. It is crucial that any rational design of coating for fiber-based e-textiles be based on the whole gamut of chemical, physical, and mechanical properties of the coating material and its compatibility with the human environment.

Author Contributions: Conceptualization, K.C. and T.K.G.; project administration, K.C. and T.K.G.; writing—original draft, K.C. and J.T.; Writing—review and editing, K.C., J.T. and T.K.G.

Funding: This review was funded by NCSU CHANCELLOR'S INNOVATION FUND, and NSFSCH grant number 1622451.

Conflicts of Interest: The authors declare no conflict of interest.

References

1. Kapoor, A.; McKnight, M.; Chatterjee, K.; Agcayazi, T.; Kausche, H.; Bozkurt, A.; Ghosh, T.K. Toward Fully Manufacturable, Fiber Assembly-Based Concurrent Multimodal and Multifunctional Sensors for e-Textiles. *Adv. Mater. Technol.* **2019**, *4*, 1800281. [CrossRef]
2. Frutiger, A.; Muth, J.T.; Vogt, D.M.; Menguc, Y.; Campo, A.; Valentine, A.D.; Walsh, C.J.; Lewis, J.A. Capacitive soft strain sensors via multicore-shell fiber printing. *Adv. Mater.* **2015**, *27*, 2440–2446. [CrossRef] [PubMed]
3. Huang, C.; Shen, C.; Tang, C.; Chang, S. A wearable yarn-based piezo-resistive sensor. *Sens. Actuators A Phys.* **2008**, *141*, 396–403. [CrossRef]

4. Oh, J.Y.; Rondeau-Gagne, S.; Chiu, Y.; Chortos, A.; Lissel, F.; Wang, G.N.; Schroeder, B.C.; Kurosawa, T.; Lopez, J.; Katsumata, T.; et al. Intrinsically stretchable and healable semiconducting polymer for organic transistors. *Nature* **2016**, *539*, 411–415. [CrossRef] [PubMed]
5. Xu, J.; Wang, S.; Wang, G.N.; Zhu, C.; Luo, S.; Jin, L.; Gu, X.; Chen, S.; Feig, V.R.; To, J.W.F.; et al. Highly stretchable polymer semiconductor films through the nanoconfinement effect. *Science* **2017**, *355*, 59. [CrossRef] [PubMed]
6. Maccioni, M.; Orgiu, E.; Cosseddu, P.; Locci, S.; Bonfiglio, A. Towards the textile transistor: Assembly and characterization of an organic field effect transistor with a cylindrical geometry. *Appl. Phys. Lett.* **2006**, *89*, 143515. [CrossRef]
7. Bonfiglio, A.; de Rossi, D.; Kirstein, T.; Locher, I.R.; Mameli, F.; Paradiso, R.; Vozzi, G. Organic field effect transistors for textile applications. *IEEE Trans. Inf. Technol. Biomed.* **2005**, *9*, 319–324. [CrossRef]
8. Jung, S.; Lauterbach, C.; Strasser, M.; Weber, W. Enabling technologies for disappearing electronics in smart textiles. In Proceedings of the 2003 IEEE International Solid-State Circuits Conference, San Francisco, CA, USA, 13 February 2003; pp. 386–387.
9. Catrysse, M.; Puers, R.; Hertleer, C.; Van Langenhove, L.; Van Egmond, H.; Matthys, D. Towards the integration of textile sensors in a wireless monitoring suit. *Sens. Actuators A Phys.* **2004**, *114*, 302–311. [CrossRef]
10. Zhu, S.; Langley, R. Dual-Band Wearable Textile Antenna on an EBG Substrate. *TAP* **2009**, *57*, 926–935. [CrossRef]
11. Huang, Y.; Hu, H.; Huang, Y.; Zhu, M.; Meng, W.; Liu, C.; Pei, Z.; Hao, C.; Wang, Z.; Zhi, C. From Industrially Weavable and Knittable Highly Conductive Yarns to Large Wearable Energy Storage Textiles. *ACS Nano* **2015**, *9*, 4766–4775. [CrossRef]
12. Soin, N.; Shah, T.H.; Anand, S.C.; Geng, J.; Pornwannachai, W.; Mandal, P.; Reid, D.; Sharma, S.; Hadimani, R.L.; Bayramol, D.V.; et al. Novel "3-D spacer" all fibre piezoelectric textiles for energy harvesting applications. *Energy Environ. Sci.* **2014**, *7*, 1670–1679. [CrossRef]
13. Kim, S.H.; Haines, C.S.; Li, N.; Kim, K.J.; Mun, T.J.; Choi, C.; Di, J.; Oh, Y.J.; Oviedo, J.P.; Bykova, J.; et al. Harvesting electrical energy from carbon nanotube yarn twist. *Science* **2017**, *357*, 773–778. [CrossRef] [PubMed]
14. Ostfeld, A.E.; Gaikwad, A.M.; Khan, Y.; Arias, A.C. High-performance flexible energy storage and harvesting system for wearable electronics. *Sci. Rep.* **2016**, *6*, 26122. [CrossRef]
15. Choe, K.; Kim, K.J.; Kim, D.; Manford, C.; Heo, S.; Shahinpoor, M. Performance Characteristics of Electro-chemically Driven Polyacrylonitrile Fiber Bundle Actuators. *J. Intell. Mater. Syst. Struct.* **2006**, *17*, 563–576. [CrossRef]
16. Di Spigna, N.; Chakraborti, P.; Yang, P.; Ghosh, T.; Franzon, P. Application of EAP materials toward a refreshable Braille display. In Proceedings of the SPIE Smart Structures and Materials + Nondestructive Evaluation and Health Monitoring, San Diego, CA, USA, 8–12 March 2009; SPIE: San Diego, CA, USA, 2009; Volume 7287, p. 9.
17. Arora, S.; Ghosh, T.; Muth, J. Dielectric elastomer based prototype fiber actuators. *Sens. Actuators A Phys.* **2007**, *136*, 321–328. [CrossRef]
18. French, J.D.; Weitz, G.E.; Luke, J.E.; Cass, R.B.; Jadidian, B.; Janas, V.; Safari, A. Production of continuous piezoelectric fibers for sensor/actuator applications. In Proceedings of the ISAF '96. Proceedings of the Tenth IEEE International Symposium on Applications of Ferroelectrics, East Brunswick, NJ, USA, 18–21 August 1996; Volume 2, p. 870.
19. Talha, A.; Kony, C.; Alper, B.; Ghosh, T.K. Flexible Interconnects for Electronic Textiles. *Adv. Mater. Technol.* **2018**, *3*, 1700277.
20. Hamedi, M.; Forchheimer, R.; Inganas, O. Towards woven logic from organic electronic fibres. *Nat. Mater.* **2007**, *6*, 357–362. [CrossRef]
21. Pang, C.; Lee, G.; Kim, T.; Kim, S.M.; Kim, H.N.; Ahn, S.; Suh, K. A flexible and highly sensitive strain-gauge sensor using reversible interlocking of nanofibres. *Nat. Mater.* **2012**, *11*, 795–801. [CrossRef]
22. Seyedin, S.; Razal, J.M.; Innis, P.C.; Jeiranikhameneh, A.; Beirne, S.; Wallace, G.G. Knitted strain sensor textiles of highly conductive all-polymeric fibers. *ACS Appl. Mater. Interfaces* **2015**, *7*, 21150–21158. [CrossRef]
23. Bashir, T.; Skrifvars, M.; Persson, N. Synthesis of high performance, conductive PEDOT-coated polyester yarns by OCVD technique. *Polym. Adv. Technol.* **2012**, *23*, 611–617. [CrossRef]

24. Lee, J.; Kwon, H.; Seo, J.; Shin, S.; Koo, J.H.; Pang, C.; Son, S.; Kim, J.H.; Jang, Y.H.; Kim, D.E.; et al. Conductive Fiber-Based Ultrasensitive Textile Pressure Sensor for Wearable Electronics. *Adv. Mater.* **2015**, *27*, 2433–2439. [CrossRef] [PubMed]
25. Sekitani, T.; Noguchi, Y.; Hata, K.; Fukushima, T.; Aida, T.; Someya, T. A Rubberlike Stretchable Active Matrix Using Elastic Conductors. *Science* **2008**, *321*, 1468. [CrossRef] [PubMed]
26. Lee, J.; Yoon, J.; Kim, H.G.; Kang, S.; Oh, W.; Algadi, H.; Al-sayari, S.; Shong, B.; Kim, S.; Kim, H.; et al. Highly conductive and flexible fiber for textile electronics obtained by extremely low-temperature atomic layer deposition of Pt. *NPG Asia Mater.* **2016**, *8*, e331. [CrossRef]
27. Lipomi, D.J.; Vosgueritchian, M.; Tee, B.C.; Hellstrom, S.L.; Lee, J.A.; Fox, C.H.; Bao, Z. Skin-like pressure and strain sensors based on transparent elastic films of carbon nanotubes. *Nat. Nanotechnol.* **2011**, *6*, 788–792. [CrossRef]
28. Yamada, T.; Hayamizu, Y.; Yamamoto, Y.; Yomogida, Y.; Izadi-Najafabadi, A.; Futaba, D.N.; Hata, K. A stretchable carbon nanotube strain sensor for human-motion detection. *Nat. Nanotechnol.* **2011**, *6*, 296. [CrossRef] [PubMed]
29. Yun, Y.J.; Hong, W.G.; Choi, N.; Kim, B.H.; Jun, Y.; Lee, H. Ultrasensitive and Highly Selective Graphene-Based Single Yarn for Use in Wearable Gas Sensor. *Sci. Rep.* **2015**, *5*, 10904. [CrossRef]
30. Neves, A.I.S.; Bointon, T.H.; Melo, L.V.; Russo, S.; de Schrijver, I.; Craciun, M.F.; Alves, H. Transparent conductive graphene textile fibers. *Sci. Rep.* **2015**, *5*, 9866. [CrossRef]
31. Zhao, Z.; Yan, C.; Liu, Z.; Fu, X.; Peng, L.; Hu, Y.; Zheng, Z. Machine-Washable Textile Triboelectric Nanogenerators for Effective Human Respiratory Monitoring through Loom Weaving of Metallic Yarns. *Adv. Mater.* **2016**, *28*, 10267–10274. [CrossRef]
32. Huang, Y.; Ip, W.S.; Lau, Y.Y.; Sun, J.; Zeng, J.; Yeung, N.S.S.; Ng, W.S.; Li, H.; Pei, Z.; Xue, Q.; et al. Weavable, Conductive Yarn-Based NiCo//Zn Textile Battery with High Energy Density and Rate Capability. *ACS Nano* **2017**, *11*, 8953–8961. [CrossRef]
33. Schwarz, A.; Hakuzimana, J.; Kaczynska, A.; Banaszczyk, J.; Westbroek, P.; McAdams, E.; Moody, G.; Chronis, Y.; Priniotakis, G.; De Mey, G.; et al. Gold coated para-aramid yarns through electroless deposition. *Surf. Coat. Technol.* **2010**, *204*, 1412–1418. [CrossRef]
34. Hu, L.; Pasta, M.; Mantia, F.L.; Cui, L.; Jeong, S.; Deshazer, H.D.; Choi, J.W.; Han, S.M.; Cui, Y. Stretchable, Porous, and Conductive Energy Textiles. *Nano Lett.* **2010**, *10*, 708–714. [CrossRef] [PubMed]
35. Shim, B.S.; Chen, W.; Doty, C.; Xu, C.; Kotov, N.A. Smart Electronic Yarns and Wearable Fabrics for Human Biomonitoring made by Carbon Nanotube Coating with Polyelectrolytes. *Nano Lett.* **2008**, *8*, 4151–4157. [CrossRef] [PubMed]
36. Cherenack, K.; Zysset, C.; Kinkeldei, T.; Münzenrieder, N.; Tröster, G. Woven Electronic Fibers with Sensing and Display Functions for Smart Textiles. *Adv. Mater.* **2010**, *22*, 5178–5182. [CrossRef] [PubMed]
37. Drabik, M.; Vogel-Schäuble, N.; Heuberger, M.; Hegemann, D.; Biederman, H. Sensors on Textile Fibres Based on Ag/a-C:H:O Nanocomposite Coatings. *Nanomater. Nanotechnol.* **2013**, *3*, 13. [CrossRef]
38. Neves, A.I.S.; Rodrigues, D.P.; De Sanctis, A.; Alonso, E.T.; Pereira, M.S.; Amaral, V.S.; Melo, L.V.; Russo, S.; de Schrijver, I.; Alves, H.; et al. Towards conductive textiles: Coating polymeric fibres with graphene. *Sci. Rep.* **2017**, *7*, 4250. [CrossRef] [PubMed]
39. Kim, B.; Koncar, V.; Dufour, C. Polyaniline-coated PET conductive yarns: Study of electrical, mechanical, and electro-mechanical properties. *J. Appl. Polym. Sci.* **2006**, *101*, 1252–1256. [CrossRef]
40. Shirakawa, H.; Louis, E.J.; MacDiarmid, A.G.; Chiang, C.K.; Heeger, A.J. Synthesis of electrically conducting organic polymers: Halogen derivatives of polyacetylene, (CH)x. *J. Chem. Soc. Chem. Commun.* **1977**, 578–580. [CrossRef]
41. Stenger-Smith, J.D. Intrinsically electrically conducting polymers. Synthesis, characterization, and their applications. *Prog. Polym. Sci.* **1998**, *23*, 57–79. [CrossRef]
42. Naveen, M.H.; Gurudatt, N.G.; Shim, Y. Applications of conducting polymer composites to electrochemical sensors: A review. *Appl. Mater. Today* **2017**, *9*, 419–433. [CrossRef]
43. Reynolds, J.R.; Skotheim, T.A. *Handbook of Conducting Polymers, 2 Volume Set*, 3rd ed.; Skotheim, T.A., Reynolds, J., Eds.; CRC Press: Boca Raton, FL, USA, 2007.
44. Ala, O.; Fan, Q. Applications of Conducting Polymers in Electronic Textiles. *Res. J. Text. Appar.* **2009**, *13*, 51–68. [CrossRef]
45. Gregory, R.V.; Kimbrell, W.C.; Kuhn, H.H. Conductive textiles. *Synth. Met.* **1989**, *28*, 823–835. [CrossRef]

46. Allison, L.; Hoxie, S.; Andrew, T.L. Towards seamlessly-integrated textile electronics: Methods to coat fabrics and fibers with conducting polymers for electronic applications. *Chem. Commun.* **2017**, *53*, 7182–7193. [CrossRef] [PubMed]
47. Kim, B.; Koncar, V.; Devaux, E.; Dufour, C.; Viallier, P. Electrical and morphological properties of PP and PET conductive polymer fibers. *Synth. Met.* **2004**, *146*, 167–174. [CrossRef]
48. Mostafalu, P.; Akbari, M.; Alberti, K.A.; Xu, Q.; Khademhosseini, A.; Sonkusale, S.R. A toolkit of thread-based microfluidics, sensors, and electronics for 3D tissue embedding for medical diagnostics. *Microsyst. Nanoeng.* **2016**, *2*, 16039. [CrossRef] [PubMed]
49. Wallace, G.; Campbell, T.; Innis, P. Putting function into fashion: Organic conducting polymer fibres and textiles. *Fibers Polym.* **2007**, *8*, 135–142. [CrossRef]
50. Yue, B.; Wang, C.; Ding, X.; Wallace, G.G. Polypyrrole coated nylon lycra fabric as stretchable electrode for supercapacitor applications. *Electrochim. Acta* **2012**, *68*, 18–24. [CrossRef]
51. Sarvi, A.; Silva, A.B.; Bretas, R.E.; Sundararaj, U. A new approach for conductive network formation in electrospun poly(vinylidene fluoride) nanofibers. *Polym. Int.* **2015**, *64*, 1262–1267. [CrossRef]
52. Eom, J.; Jaisutti, R.; Lee, H.; Lee, W.; Heo, J.; Lee, J.; Park, S.K.; Kim, Y. Highly Sensitive Textile Strain Sensors and Wireless User-Interface Devices Using All-Polymeric Conducting Fibers. *ACS Appl. Mater. Interfaces* **2017**, *9*, 10190–10197. [CrossRef] [PubMed]
53. Hong, K.H.; Oh, K.W.; Kang, T.J. Preparation and properties of electrically conducting textiles by in situ polymerization of poly(3,4-ethylenedioxythiophene). *J. Appl. Polym. Sci.* **2005**, *97*, 1326–1332. [CrossRef]
54. Zhang, L.; Fairbanks, M.; Andrew, T.L. Rugged Textile Electrodes for Wearable Devices Obtained by Vapor Coating Off-the-Shelf, Plain-Woven Fabrics. *Adv. Funct. Mater.* **2017**, *27*, 1700415. [CrossRef]
55. Alf, M.E.; Asatekin, A.; Barr, M.C.; Baxamusa, S.H.; Chelawat, H.; Ozaydin-Ince, G.; Petruczok, C.D.; Sreenivasan, R.; Tenhaeff, W.E.; Trujillo, N.J.; et al. Chemical Vapor Deposition of Conformal, Functional, and Responsive Polymer Films. *Adv. Mater.* **2010**, *22*, 1993–2027. [CrossRef] [PubMed]
56. Bashir, T.; Skrifvars, M.; Persson, N. Production of highly conductive textile viscose yarns by chemical vapor deposition technique: A route to continuous process. *Polym. Adv. Technol.* **2011**, *22*, 2214–2221. [CrossRef]
57. Tenhaeff, W.E.; Gleason, K.K. Initiated and Oxidative Chemical Vapor Deposition of Polymeric Thin Films: iCVD and oCVD. *Adv. Funct. Mater.* **2008**, *18*, 979–992. [CrossRef]
58. Han, J.; Kim, B.; Li, J.; Meyyappan, M. A carbon nanotube based ammonia sensor on cotton textile. *Appl. Phys. Lett.* **2013**, *102*, 193104. [CrossRef]
59. Huang, J. Carbon black filled conducting polymers and polymer blends. *Adv. Polym. Technol.* **2002**, *21*, 299–313. [CrossRef]
60. De Volder, M.F.L.; Tawfick, S.H.; Baughman, R.H.; Hart, A.J. Carbon Nanotubes: Present and Future Commercial Applications. *Science* **2013**, *339*, 535–539. [CrossRef]
61. Aharoni, S.M. Electrical Resistivity of a Composite of Conducting Particles in an Insulating Matrix. *J. Appl. Phys.* **1972**, *43*, 2463–2465. [CrossRef]
62. Alimohammadi, F.; Parvinzadeh Gashti, M.; Shamei, A. Functional cellulose fibers via polycarboxylic acid/carbon nanotube composite coating. *J. Coat. Technol. Res.* **2013**, *10*, 123–132. [CrossRef]
63. Iijima, S. Helical microtubules of graphitic carbon. *Nature* **1991**, *354*, 56–58. [CrossRef]
64. Coleman, J.N.; Curran, S.; Dalton, A.B.; Davey, A.P.; McCarthy, B.; Blau, W.; Barklie, R.C. Percolation-dominated conductivity in a conjugated-polymer-carbon-nanotube composite. *Phys. Rev. B* **1998**, *58*, R7495. [CrossRef]
65. Bauhofer, W.; Kovacs, J.Z. A review and analysis of electrical percolation in carbon nanotube polymer composites. *Compos. Sci. Technol.* **2009**, *69*, 1486–1498. [CrossRef]
66. Zhang, R.; Deng, H.; Valenca, R.; Jin, J.; Fu, Q.; Bilotti, E.; Peijs, T. Carbon nanotube polymer coatings for textile yarns with good strain sensing capability. *Sens. Actuators A Phys.* **2012**, *179*, 83–91. [CrossRef]
67. Bai, J.B.; Allaoui, A. Effect of the length and the aggregate size of MWNTs on the improvement efficiency of the mechanical and electrical properties of nanocomposites—Experimental investigation. *Compos. Part A* **2003**, *34*, 689–694. [CrossRef]
68. Behnam, A.; Ural, A.; Guo, J. Effects of nanotube alignment and measurement direction on percolation resistivity in single-walled carbon nanotube films. *J. Appl. Phys.* **2007**, *102*, 7. [CrossRef]
69. Baughman, R.H.; Zakhidov, A.A.; De Heer, W.A. Carbon Nanotubes—The Route Toward Applications. *Science* **2002**, *297*, 787–792. [CrossRef]

70. Venema, L.C.; Dekker, C.; Smalley, R.E.; Rinzler, A.G.; Wilder, J.W.G. Electronic structure of atomically resolved carbon nanotubes. *Nature* **1998**, *391*, 59–62.
71. Thess, A.; Lee, R.; Nikolaev, P.; Dai, H.; Petit, P.; Robert, J.; Xu, C.; Lee, Y.H.; Kim, S.G.; Rinzler, A.G.; et al. Crystalline Ropes of Metallic Carbon Nanotubes. *Science* **1996**, *273*, 483–487. [CrossRef]
72. Zhao, H.; Zhang, Y.; Bradford, P.D.; Zhou, Q.; Jia, Q.; Yuan, F.; Zhu, Y. Carbon nanotube yarn strain sensors. *Nanotechnology* **2010**, *21*, 305502. [CrossRef]
73. Li, Y.; Zhou, B.; Zheng, G.; Liu, X.; Li, T.; Yan, C.; Cheng, C.; Dai, K.; Liu, C.; Shen, C.; et al. Continuously prepared highly conductive and stretchable SWNT/MWNT synergistically composited electrospun thermoplastic polyurethane yarns for wearable sensing. *J. Mater. Chem. C* **2018**, *6*, 2258–2269. [CrossRef]
74. Liu, Y.; Wang, X.; Qi, K.; Xin, J.H. Functionalization of cotton with carbon nanotubes. *J. Mater. Chem.* **2008**, *29*, 3454–3460. [CrossRef]
75. Xin, G.; Yao, T.; Sun, H.; Scott, S.M.; Shao, D.; Wang, G.; Lian, J. Highly thermally conductive and mechanically strong graphene fibers. *Science* **2015**, *349*, 1083–1087. [CrossRef] [PubMed]
76. Kotov, N.A. Carbon sheet solutions. *Nature* **2006**, *442*, 254–255. [CrossRef] [PubMed]
77. Ponnamma, D.; Guo, Q.; Krupa, I.; Al-Maadeed, M.A.S.A.; Varughese, K.T.; Thomas, S.; Sadasivuni, K.K. Graphene and graphitic derivative filled polymer composites as potential sensors. *Phys. Chem. Chem. Phys.* **2015**, *17*, 3954–3981. [CrossRef] [PubMed]
78. Li, X.; Hua, T.; Xu, B. Electromechanical properties of a yarn strain sensor with graphene-sheath/polyurethane-core. *Carbon* **2017**, *118*, 686–698. [CrossRef]
79. Novoselov, K.S.; Geim, A.K.; Morozov, S.V.; Jiang, D.; Zhang, Y.; Dubonos, S.V.; Grigorieva, I.V.; Firsov, A.A. Electric Field Effect in Atomically Thin Carbon Films. *Science* **2004**, *306*, 666–669. [CrossRef]
80. Hummers, W.S.; Offeman, R.S. Preparation of graphitic oxide. *J. Am. Chem. Soc.* **1958**, *80*, 1339. [CrossRef]
81. Stankovich, S.; Dikin, D.A.; Piner, R.D.; Kohlhaas, K.A.; Kleinhammes, A.; Jia, Y.; Wu, Y.; Nguyen, S.T.; Ruoff, R.S. Synthesis of graphene-based nanosheets via chemical reduction of exfoliated graphite oxide. *Carbon* **2007**, *45*, 1558–1565. [CrossRef]
82. Park, J.J.; Hyun, W.J.; Mun, S.C.; Park, Y.T.; Park, O.O. Highly Stretchable and Wearable Graphene Strain Sensors with Controllable Sensitivity for Human Motion Monitoring. *ACS Appl. Mater. Interfaces* **2015**, *7*, 6317–6324. [CrossRef]
83. Ji, X.; Xu, Y.; Zhang, W.; Cui, L.; Liu, J. Review of functionalization, structure and properties of graphene/polymer composite fibers. *Compos. Part A* **2016**, *87*, 29–45. [CrossRef]
84. Xiang, C.; Lu, W.; Zhu, Y.; Sun, Z.; Yan, Z.; Hwang, C.; Tour, J.M. Carbon nanotube and graphene nanoribbon-coated conductive kevlar fibers. *ACS Appl. Mater. Interfaces* **2012**, *4*, 131–136. [CrossRef]
85. Zhang, M.; Wang, C.; Wang, Q.; Jian, M.; Zhang, Y. Sheath-Core Graphite/Silk Fiber Made by Dry-Meyer-Rod-Coating for Wearable Strain Sensors. *ACS Appl. Mater. Interfaces* **2016**, *8*, 20894–20899. [CrossRef] [PubMed]
86. Yun, Y.J.; Hong, W.G.; Kim, W.; Jun, Y.; Kim, B.H. A Novel Method for Applying Reduced Graphene Oxide Directly to Electronic Textiles from Yarns to Fabrics. *Adv. Mater.* **2013**, *25*, 5701–5705. [CrossRef] [PubMed]
87. Knite, M.; Teteris, V.; Kiploka, A.; Kaupuzs, J. Polyisoprene-carbon black nanocomposites as tensile strain and pressure sensor materials. *Sens. Actuators. A Phys.* **2004**, *110*, 142–149. [CrossRef]
88. Chatterjee, A.; Alam, K.; Klein, P. Electrically Conductive Carbon Nanofiber Composites with High-Density Polyethylene and Glass Fibers. *Mater. Manuf. Process.* **2007**, *22*, 62–65. [CrossRef]
89. Al-Saleh, M.H.; Sundararaj, U. A review of vapor grown carbon nanofiber/polymer conductive composites. *Carbon* **2009**, *47*, 2–22. [CrossRef]
90. Jin, H.; Matsuhisa, N.; Lee, S.; Abbas, M.; Yokota, T.; Someya, T. Enhancing the Performance of Stretchable Conductors for E-Textiles by Controlled Ink Permeation. *Adv. Mater.* **2017**, *29*, 1605848. [CrossRef] [PubMed]
91. Nilsson, E.; Rigdahl, M.; Hagström, B. Electrically conductive polymeric bi-component fibers containing a high load of low-structured carbon black. *J. Appl. Polym. Sci.* **2015**, *132*, 42255. [CrossRef]
92. Villanueva, R.; Ganta, D.; Guzman, C. Mechanical, in-situ electrical and thermal properties of wearable conductive textile yarn coated with polypyrrole/carbon black composite. *Mater. Res.* **2018**, *6*, 16307. [CrossRef]
93. Souri, H.; Bhattacharyya, D. Wearable strain sensors based on electrically conductive natural fiber yarns. *Mater. Des.* **2018**, *154*, 217–227. [CrossRef]

94. Sabetzadeh, N.; Najar, S.S.; Bahrami, S.H. Electrical conductivity of vapor-grown carbon nanofiber/polyester textile-based composites. *J. Appl. Polym. Sci.* **2013**, *130*, 3009–3017. [CrossRef]
95. Guo, H.; Rasheed, A.; Minus, M.; Kumar, S. Polyacrylonitrile/vapor grown carbon nanofiber composite films. *J. Mater. Sci.* **2008**, *43*, 4363–4369. [CrossRef]
96. Kim, H.; Lee, S. Characterization of carbon nanofiber (CNF)/polymer composite coated on cotton fabrics prepared with various circuit patterns. *Fash. Text.* **2018**, *5*, 1–13. [CrossRef]
97. Chowdhury, S.; Olima, M.; Liu, Y.; Saha, M.; Bergman, J.; Robison, T. Poly dimethylsiloxane/carbon nanofiber nanocomposites: Fabrication and characterization of electrical and thermal properties. *Int. J. Smart Nano Mater.* **2016**, *7*, 236–247. [CrossRef]
98. Narayanan, S.; Karpagam, K.; Bhattacharyya, A. Nanocomposite coatings on cotton and silk fibers for enhanced electrical conductivity. *Fibers Polym.* **2015**, *16*, 1269–1275. [CrossRef]
99. Rodriguez, A.J.; Guzman, M.E.; Lim, C.; Minaie, B. Mechanical properties of carbon nanofiber/fiber-reinforced hierarchical polymer composites manufactured with multiscale-reinforcement fabrics. *Carbon* **2011**, *49*, 937–948. [CrossRef]
100. Duan, H.; Liang, J.; Xia, Z. Synthetic hierarchical nanostructures: Growth of carbon nanofibers on microfibers by chemical vapor deposition. *Mater. Sci. Eng. B* **2010**, *166*, 190–195. [CrossRef]
101. Little, B.K.; Li, Y.; Cammarata, V.; Broughton, R.; Mills, G. Metallization of Kevlar Fibers with Gold. *ACS Appl. Mater. Interfaces* **2011**, *3*, 1965–1973. [CrossRef]
102. Liu, X.Q.; Chang, H.X.; Li, Y.; Huck, W.T.S.; Zheng, Z.J. Polyelectrolyte-Bridged Metal/Cotton Hierarchical Structures for Highly Durable Conductive Yarns. *ACS Appl. Mater. Interfaces* **2010**, *2*, 529–535. [CrossRef]
103. Lee, H.M.; Choi, S.; Jung, A.; Ko, S.H. Highly Conductive Aluminum Textile and Paper for Flexible and Wearable Electronics. *Angew. Chem. Int. Ed. Engl.* **2013**, *52*, 7718–7723. [CrossRef]
104. Gamburg, Y.; Zangari, G. Introduction to Electrodeposition: Basic Terms and Fundamental Concepts. In *Theory and Practice of Metal Electrodeposition*; Springer: New York, NY, USA, 2011; pp. 1–25.
105. Dini, J.W.; Hajdu, J.B. *Electrodeposition: The Materials Science of Coatings and Substrates*; William Andrew Publishing: Norwich, UK, 1993.
106. Mallory, G.O.; Hajdu, J.B. *Electroless Plating—Fundamentals and Applications*; William Andrew Publishing: New York, NY, USA, 1990.
107. Jur, J.S.; Sweet, W.J.; Oldham, C.J.; Parsons, G.N. Atomic Layer Deposition of Conductive Coatings on Cotton, Paper, and Synthetic Fibers: Conductivity Analysis and Functional Chemical Sensing Using All-Fiber Capacitors. *Adv. Funct. Mater.* **2011**, *21*, 1993–2002. [CrossRef]
108. Park, H.J.; Kim, W.; Ah, C.S.; Jun, Y.; Yun, Y.J. Solution-processed Au–Ag core–shell nanoparticle-decorated yarns for human motion monitoring. *RSC Adv.* **2017**, *7*, 10539–10544. [CrossRef]
109. Husain, M.D.; Kennon, R.; Dias, T. Design and fabrication of Temperature Sensing Fabric. *J. Ind. Text.* **2014**, *44*, 398–417. [CrossRef]
110. Shang, Y.; Li, Y.; He, X.; Du, S.; Zhang, L.; Shi, E.; Wu, S.; Li, Z.; Li, P.; Wei, J.; et al. Highly Twisted Double-Helix Carbon Nanotube Yarns. *ACS Nano* **2013**, *7*, 1446–1453. [CrossRef] [PubMed]
111. Seesaard, T.; Lorwongtragool, P.; Kerdcharoen, T. Development of Fabric-Based Chemical Gas Sensors for Use as Wearable Electronic Noses. *Sensors* **2015**, *15*, 1885–1902. [CrossRef]
112. Wang, H.; Liu, Z.; Ding, J.; Lepró, X.; Fang, S.; Jiang, N.; Yuan, N.; Wang, R.; Yin, Q.; Lv, W.; et al. Downsized Sheath–Core Conducting Fibers for Weavable Superelastic Wires, Biosensors, Supercapacitors, and Strain Sensors. *Adv. Mater.* **2016**, *28*, 4998–5007. [CrossRef]
113. Choi, C.; Lee, J.M.; Kim, S.H.; Kim, S.J.; Di, J.; Baughman, R.H. Twistable and Stretchable Sandwich Structured Fiber for Wearable Sensors and Supercapacitors. *Nano Lett.* **2016**, *16*, 7677–7684. [CrossRef] [PubMed]
114. Wei, Y.; Chen, S.; Yuan, X.; Wang, P.; Liu, L. Multiscale Wrinkled Microstructures for Piezoresistive Fibers. *Adv. Funct. Mater.* **2016**, *26*, 5078–5085. [CrossRef]
115. Chen, S.; Lou, Z.; Chen, D.; Jiang, K.; Shen, G. Polymer-Enhanced Highly Stretchable Conductive Fiber Strain Sensor Used for Electronic Data Gloves. *Adv. Mater. Technol.* **2016**, *1*, 1600136. [CrossRef]
116. Cheng, Y.; Wang, R.; Sun, J.; Gao, L. A Stretchable and Highly Sensitive Graphene-Based Fiber for Sensing Tensile Strain, Bending, and Torsion. *Adv. Mater.* **2015**, *27*, 7365–7371. [CrossRef]
117. Tai, Y.; Lubineau, G. Double-Twisted Conductive Smart Threads Comprising a Homogeneously and a Gradient-Coated Thread for Multidimensional Flexible Pressure-Sensing Devices. *Adv. Funct. Mater.* **2016**, *26*, 4078–4084. [CrossRef]

118. Cheng, Y.; Wang, R.; Zhai, H.; Sun, J. Stretchable electronic skin based on silver nanowire composite fiber electrodes for sensing pressure, proximity, and multidirectional strain. *Nanoscale* **2017**, *9*, 3834–3842. [CrossRef] [PubMed]
119. Chhetry, A.; Yoon, H.; Park, J.Y. A flexible and highly sensitive capacitive pressure sensor based on conductive fibers with a microporous dielectric for wearable electronics. *J. Mater. Chem. C* **2017**, *5*, 10068–10076. [CrossRef]
120. Harnett, C.K.; Zhao, H.; Shepherd, R.F. Stretchable Optical Fibers: Threads for Strain-Sensitive Textiles. *Adv. Mater. Technol.* **2017**, *2*, 1700087. [CrossRef]
121. Guo, J.; Liu, X.; Jiang, N.; Yetisen, A.K.; Yuk, H.; Yang, C.; Khademhosseini, A.; Zhao, X.; Yun, S. Highly Stretchable, Strain Sensing Hydrogel Optical Fibers. *Adv. Mater.* **2016**, *28*, 10244–10249. [CrossRef] [PubMed]
122. Zhang, Y.; Feng, D.; Liu, Z.; Guo, Z.; Dong, X.; Chiang, K.S.; Chu, B.C.B. High-sensitivity pressure sensor using a shielded polymer-coated fiber Bragg grating. *LPT* **2001**, *13*, 618–619. [CrossRef]
123. Rothmaier, M.; Luong, T.M.; Clemens, F. Textile pressure sensor made of flexible plastic optical fibers. *Sensor* **2008**, *55*, 4318–4329. [CrossRef]
124. Liao, Q.; Mohr, M.; Zhang, X.; Zhang, Z.; Zhang, Y.; Fecht, H. Carbon fiber-ZnO nanowire hybrid structures for flexible and adaptable strain sensors. *Nanoscale* **2013**, *5*, 12350. [CrossRef] [PubMed]
125. Li, X.; Lin, Z.; Cheng, G.; Wen, X.; Liu, Y.; Niu, S.; Wang, Z.L. 3D Fiber-Based Hybrid Nanogenerator for Energy Harvesting and as a Self-Powered Pressure Sensor. *ACS Nano* **2014**, *8*, 10674–10681. [CrossRef]
126. Ge, J.; Sun, L.; Zhang, F.; Zhang, Y.; Shi, L.; Zhao, H.; Zhu, H.; Jiang, H.; Yu, S. A Stretchable Electronic Fabric Artificial Skin with Pressure-, Lateral Strain-, and Flexion-Sensitive Properties. *Adv. Mater.* **2016**, *28*, 722–728. [CrossRef]
127. Hasegawa, Y.; Shikida, M.; Ogura, D.; Suzuki, Y.; Sato, K. Fabrication of a wearable fabric tactile sensor produced by artificial hollow fiber. *J. Micromech. Microeng.* **2008**, *18*, 085014. [CrossRef]
128. Atalay, O.; Kennon, W.R. Knitted Strain Sensors: Impact of Design Parameters on Sensing Properties. *Sensors* **2014**, *14*, 4712–4730. [CrossRef] [PubMed]
129. Seyedin, S.; Zhang, P.; Naebe, M.; Qin, S.; Chen, J.; Wang, X.; Razal, J.M. Textile strain sensors: A review of the fabrication technologies, performance evaluation and applications. *Mater. Horiz.* **2019**, *6*, 219–249. [CrossRef]
130. Fan, Q.; Zhang, X.; Qin, Z. Preparation of Polyaniline/Polyurethane Fibers and Their Piezoresistive Property. *J. Macromol. Sci. B* **2012**, *51*, 736–746. [CrossRef]
131. Kloeck, B.; De Rooij, N.F. *Mechanical Sensors*; Sze, S.M., Ed.; Semiconductor Sensors; John Wiley & Sons, Inc.: New York, NY, USA, 1994; pp. 153–204.
132. Wu, H.; Liu, Q.; Chen, H.; Shi, G.; Li, C. Fibrous strain sensor with ultra-sensitivity, wide sensing range, and large linearity for full-range detection of human motion. *Nanoscale* **2018**, *10*, 17512–17519. [CrossRef] [PubMed]
133. Kang, T.J.; Choi, A.; Kim, D.; Jin, K.; Seo, D.K.; Jeong, D.H.; Hong, S.; Park, Y.W.; Kim, Y.H. Electromechanical properties of CNT-coated cotton yarn for electronic textile applications. *Smart Mater. Struct.* **2011**, *20*, 015004. [CrossRef]
134. Huang, Y.; Kershaw, S.V.; Wang, Z.; Pei, Z.; Liu, J.; Huang, Y.; Li, H.; Zhu, M.; Rogach, A.L.; Zhi, C. Highly Integrated Supercapacitor-Sensor Systems via Material and Geometry Design. *Small* **2016**, *12*, 3393–3399. [CrossRef]
135. Arthur, V.H. *Dielectric Materials and Applications*; Artech House: Boston, MA, USA, 1954.
136. Takamatsu, S.; Kobayashi, T.; Shibayama, N.; Miyake, K.; Itoh, T. Fabric pressure sensor array fabricated with die-coating and weaving techniques. *Sens. Actuators A Phys.* **2012**, *184*, 57–63. [CrossRef]
137. Yang, Z.; Deng, J.; Chen, X.; Ren, J.; Peng, H. A Highly Stretchable, Fiber-Shaped Supercapacitor. *Angew. Chem. Int. Ed.* **2013**, *52*, 13453–13457. [CrossRef]
138. Sze, S.M. *Semiconductor Sensors*; Wiley: New York, NY, USA, 1994.
139. Wei, Y.; Chen, S.; Li, F.; Lin, Y.; Zhang, Y.; Liu, L. Highly Stable and Sensitive Paper-Based Bending Sensor Using Silver Nanowires/Layered Double Hydroxides Hybrids. *ACS Appl. Mater. Interfaces* **2015**, *7*, 14182–14191. [CrossRef]
140. Dellon, E.S.; Mourey, R.; Dellon, A.L. Human Pressure Perception Values for Constant and Moving One- and Two-Point Discrimination. *Plast. Reconstr. Surg.* **1992**, *90*, 112–117. [CrossRef]

141. Mannsfeld, S.C.B.; Tee, B.C.; Stoltenberg, R.M.; Chen, C.V.H.-H.; Barman, S.; Muir, B.V.O.; Sokolov, A.N.; Reese, C.; Bao, Z. Highly sensitive flexible pressure sensors with microstructured rubber dielectric layers. *Nat. Mater.* **2010**, *9*, 859–864. [CrossRef] [PubMed]
142. Barlian, A.A.; Park, W.; Mallon, J.R.; Rastegar, A.J.; Pruitt, B.L. Review: Semiconductor Piezoresistance for Microsystems. *JPROC* **2009**, *97*, 513–552. [CrossRef] [PubMed]
143. Rowlands, A.V.; Stone, M.R.; Eston, R.G. Influence of Speed and Step Frequency during Walking and Running on Motion Sensor Output. *Med. Sci. Sports Exerc.* **2007**, *39*, 716–727. [CrossRef] [PubMed]
144. Dillman, C.J. Kinematic Analyses of Running. *Exerc. Sport Sci. Rev.* **1975**, *3*, 193–218. [CrossRef] [PubMed]
145. Qi, K.; He, J.; Wang, H.; Zhou, Y.; You, X.; Nan, N.; Shao, W.; Wang, L.; Ding, B.; Cui, S. A Highly Stretchable Nanofiber-Based Electronic Skin with Pressure-, Strain-, and Flexion-Sensitive Properties for Health and Motion Monitoring. *ACS Appl. Mater. Interfaces* **2017**, *9*, 42951–42960. [CrossRef] [PubMed]
146. Wang, C.; Li, X.; Gao, E.; Jian, M.; Xia, K.; Wang, Q.; Xu, Z.; Ren, T.; Zhang, Y. Carbonized Silk Fabric for Ultrastretchable, Highly Sensitive, and Wearable Strain Sensors. *Adv. Mater.* **2016**, *28*, 6640–6648. [CrossRef] [PubMed]
147. Fu, W.; Dai, Y.; Meng, X.; Xu, W.; Zhou, J.; Liu, Z.; Lu, W.; Wang, S.; Huang, C.; Sun, Y. Electronic textiles based on aligned electrospun belt-like cellulose acetate nanofibers and graphene sheets: Portable, scalable and eco-friendly strain sensor. *Nanotechnology* **2019**, *30*, 045602. [CrossRef]
148. Ren, J.; Wang, C.; Zhang, X.; Carey, T.; Chen, K.; Yin, Y.; Torrisi, F. Environmentally-friendly conductive cotton fabric as flexible strain sensor based on hot press reduced graphene oxide. *Carbon* **2017**, *111*, 622–630. [CrossRef]
149. Jeon, H.; Hong, S.K.; Cho, S.J.; Lim, G. Fabrication of a Highly Sensitive Stretchable Strain Sensor Utilizing a Microfibrous Membrane and a Cracking Structure on Conducting Polymer. *Macromol. Mater. Eng.* **2018**, *303*, 1700389. [CrossRef]
150. Yang, Z.; Pang, Y.; Han, X.; Yang, Y.; Ling, J.; Jian, M.; Zhang, Y.; Yang, Y.; Ren, T. Graphene Textile Strain Sensor with Negative Resistance Variation for Human Motion Detection. *ACS Nano* **2018**, *12*, 9134–9141. [CrossRef]
151. Wang, Y.; Wang, L.; Yang, T.; Li, X.; Zang, X.; Zhu, M.; Wang, K.; Wu, D.; Zhu, H. Wearable and Highly Sensitive Graphene Strain Sensors for Human Motion Monitoring. *Adv. Funct. Mater.* **2014**, *24*, 4666–4670. [CrossRef]
152. Kim, I.; Shahariar, H.; Ingram, W.F.; Zhou, Y.; Jur, J.S. Inkjet Process for Conductive Patterning on Textiles: Maintaining Inherent Stretchability and Breathability in Knit Structures. *Adv. Funct. Mater.* **2018**, *29*, 1807573. [CrossRef]
153. Wang, Z.; Huang, Y.; Sun, J.; Huang, Y.; Hu, H.; Jiang, R.; Gai, W.; Li, G.; Zhi, C. Polyurethane/Cotton/Carbon Nanotubes Core-Spun Yarn as High Reliability Stretchable Strain Sensor for Human Motion Detection. *ACS Appl. Mater. Interfaces* **2016**, *8*, 24837–24843. [CrossRef] [PubMed]
154. Hao, B.; Ma, P.; Ma, Q.; Yang, S.; Mäder, E. Comparative study on monitoring structural damage in fiber-reinforced polymers using glass fibers with carbon nanotubes and graphene coating. *Compos. Sci. Technol.* **2016**, *129*, 38–45. [CrossRef]
155. Zhang, J.; Liu, J.; Zhuang, R.; Mäder, E.; Heinrich, G.; Gao, S. Single MWNT-Glass Fiber as Strain Sensor and Switch. *Adv. Mater.* **2011**, *23*, 3392–3397. [CrossRef] [PubMed]
156. Wu, X.; Han, Y.; Zhang, X.; Lu, C. Highly Sensitive, Stretchable, and Wash-Durable Strain Sensor Based on Ultrathin Conductive Layer@Polyurethane Yarn for Tiny Motion Monitoring. *ACS Appl. Mater. Interfaces* **2016**, *8*, 9936–9945. [CrossRef]
157. Kim, J.; Lee, M.; Shim, H.J.; Ghaffari, R.; Cho, H.R.; Son, D.; Jung, Y.H.; Soh, M.; Choi, C.; Jung, S.; et al. Stretchable silicon nanoribbon electronics for skin prosthesis. *Nat. Commun.* **2014**, *5*, 5747. [CrossRef]
158. Jeong, C.K.; Lee, J.; Han, S.; Ryu, J.; Hwang, G.; Park, D.Y.; Park, J.H.; Lee, S.S.; Byun, M.; Ko, S.H.; et al. A Hyper-Stretchable Elastic-Composite Energy Harvester. *Adv. Mater.* **2015**, *27*, 2866–2875. [CrossRef]
159. Abdul Samad, Y.; Komatsu, K.; Yamashita, D.; Li, Y.; Zheng, L.; Alhassan, S.M.; Nakano, Y.; Liao, K. From sewing thread to sensor: Nylon® fiber strain and pressure sensors. *Sens. Actuators. B Chem.* **2017**, *240*, 1083–1090. [CrossRef]
160. Li, J.; Xu, B. Novel highly sensitive and wearable pressure sensors from conductive three-dimensional fabric structures. *Smart Mater. Struct.* **2015**, *24*, 125022. [CrossRef]

161. Cao, R.; Pu, X.; Du, X.; Yang, W.; Wang, J.; Guo, H.; Zhao, S.; Yuan, Z.; Zhang, C.; Li, C.; et al. Screen-Printed Washable Electronic Textiles as Self-Powered Touch/Gesture Tribo-Sensors for Intelligent Human–Machine Interaction. *ACS Nano* **2018**, *12*, 5190–5196. [CrossRef] [PubMed]
162. Doshi, S.M.; Thostenson, E.T. Thin and Flexible Carbon Nanotube-Based Pressure Sensors with Ultrawide Sensing Range. *ACS Sens.* **2018**, *3*, 1276–1282. [CrossRef] [PubMed]
163. Kweon, O.Y.; Lee, S.J.; Oh, J.H. Wearable high-performance pressure sensors based on three-dimensional electrospun conductive nanofibers. *NPG Asia Mater.* **2018**, *10*, 540–551. [CrossRef]
164. Wei, Y.; Chen, S.; Dong, X.; Lin, Y.; Liu, L. Flexible piezoresistive sensors based on "dynamic bridging effect" of silver nanowires toward graphene. *Carbon* **2017**, *113*, 395. [CrossRef]
165. Liu, M.; Pu, X.; Jiang, C.; Liu, T.; Huang, X.; Chen, L.; Du, C.; Sun, J.; Hu, W.; Wang, Z.L. Large-Area All-Textile Pressure Sensors for Monitoring Human Motion and Physiological Signals. *Adv. Mater.* **2017**, *29*, 1703700. [CrossRef]

© 2019 by the authors. Licensee MDPI, Basel, Switzerland. This article is an open access article distributed under the terms and conditions of the Creative Commons Attribution (CC BY) license (http://creativecommons.org/licenses/by/4.0/).

Review

Actuator Materials: Review on Recent Advances and Future Outlook for Smart Textiles

Dharshika Kongahage and Javad Foroughi *

Intelligent Polymer Research Institute, University of Wollongong, Australia, Wollongong, NSW 2522, Australia; dharshikak@yahoo.com
* Correspondence: foroughi@uow.edu.au; Tel.: +61-242981452

Received: 24 January 2019; Accepted: 2 March 2019; Published: 11 March 2019

Abstract: Smart textiles based on actuator materials are of practical interest, but few types have been commercially exploited. The challenge for researchers has been to bring the concept out of the laboratory by working out how to build these smart materials on an industrial scale and permanently incorporate them into textiles. Smart textiles are considered as the next frontline for electronics. Recent developments in advance technologies have led to the appearance of wearable electronics by fabricating, miniaturizing and embedding flexible conductive materials into textiles. The combination of textiles and smart materials have contributed to the development of new capabilities in fabrics with the potential to change how athletes, patients, soldiers, first responders, and everyday consumers interact with their clothes and other textile products. Actuating textiles in particular, have the potential to provide a breakthrough to the area of smart textiles in many ways. The incorporation of actuating materials in to textiles is a striking approach as a small change in material anisotropy properties can be converted into significant performance enhancements, due to the densely interconnected structures. Herein, the most recent advances in smart materials based on actuating textiles are reviewed. The use of novel emerging twisted synthetic yarns, conducting polymers, hybrid carbon nanotube and spandex yarn actuators, as well as most of the cutting–edge polymeric actuators which are deployed as smart textiles are discussed.

Keywords: smart textiles; actuator; wearable technology; carbon nanotubes; conducting polymers; polymer actuators

---

1. Introduction

Smart textiles research represents an innovative model for integrating advanced engineering materials into textiles which will result in new discoveries. Smart textiles are defined as the "textiles that can sense or react to environmental conditions or stimuli, from mechanical, thermal, magnetic, chemical, electrical, or other sources in a predetermined way" [1–3]. As a more straightforward definition, textiles which can perform additional functionalities than the conventional textiles are described as smart textiles. Smart textiles have been used in numerous applications in the healthcare industry, military, and as wearable electronics [4–7]. Moreover, smart textiles can be divided in to three categories; passive, active and very smart textiles [1,8–10]. The passive smart textile is the first category of smart textiles that can provide additional features in a passive mode, irrespective of the change in the environment. As examples, anti-microbial, anti-odor, anti-static and bullet proof textiles are considered to be passive smart textiles [1]. Active smart textiles are a group that can sense and react to stimuli from the environment. These materials may also be used as sensors and actuators [1]. Very smart textiles are the third category that consists of a unit for recognizing, reasoning and actuating. This type of textiles sense, react and adapt themselves to environmental conditions or stimuli, such as space suits and health monitoring systems [11]. Textiles which can find prospective applications in energy conversion are important to smart textiles in many ways. Actuators are considered as a group which can accomplish the conversion of energy to mechanical form with the capability of moving or controlling a mechanism or a system. Actuators can reversibly contract, expand, or rotate themselves, due to the presence of an external stimulus, such as voltage, current, temperature, pressure and many more. These materials can be divided into four major groups depending on their mode of actuation which are electric field, ion based, pneumatic and thermal actuation, and then further into two major groups on whether volume or order change dominates [12]. There are several frontier actuating materials being introduced by researchers, such as carbon nanotubes, conducting polymers, and shape memory alloys [13–18]. Using actuating materials in smart textiles is an impressive approach as a small change in material properties can be converted into significant movements, due to the densely interconnected structures.

The research reported to date on actuating textiles has attempted in force/strain amplifications and to incorporate smart functionalities into fabrics. Some of the polymer actuators exhibit properties, such as the long length, high tensile strength, flexibility and durability which are essential parameters for textile yarns [17]. In addition, polymer fibers have already been used in the textile fabrication process. Therefore, the feasibility of a textile structure can be established with polymer fiber actuators. Integration of actuators into the textiles was performed in most studies using traditional textile fabrication methods, such as weaving, knitting and braiding [13,17,19]. Consequently, materials and fabrication processes for an actuating textile should be selected with careful consideration for optimum performance. This paper is mostly focused on critically reviewing and appraising the materials and processes required to fabricate a high-performance actuating textile. This review further discusses fundamental actuation mechanisms in brief, material fabrication, properties and actuating materials already being trailed in textiles.

2. Overview of Different Actuation Mechanisms

Actuator designers have introduced criteria to allow the optimal selection of actuators for a given application. Power output per mass, per volume and actuator efficiency are the three fundamental characterizing properties of actuators [20]. Furthermore, stress, strain, strain rate, cycle life and elastic modulus are some of the other general characteristics considered in the evaluation criteria [20,21]. In addition to the above technical parameters, user friendliness, ease of fabrication and maintenance, cost and availability of the raw materials are some of the additional requirements to be considered in selecting actuators for an application. It is also necessary to consider the actuation mechanism which is as important as the other performance characteristics. Most of the actuators which are being reported in the literature actuate with one of four different methods—electric field, pneumatic, ionic and thermal. This section will provide an overview of the actuation mechanisms, and their characteristics.

2.1. Electric Field Actuation.

Electric field actuation is a result of electrostatic interactions between electrodes or molecular re-organisation within the actuator material structure. These are commonly known as electronic artificial muscles and are one type of electroactive polymers (EAPs). The electric field actuation is present in low modulus polymers, such as dielectric elastomers (DEAs) and electrostrictive polymers where the electric dipoles are arranged by the electric field which result in displacement [22,23]. DEAs are simple in mechanism, construction and able to produce large strains 10% to 100% but can reach up to 380% with high electric fields [24,25]. These actuators can yield stress up to 7.7 MPa and 3.2 MPa in silicone and acrylic based actuators, respectively [26]. Due to the large strains these actuators produce high work per unit volume per cycle with a maximum of 3.4 MJ/m^3 [23].

During the actuation of DEAs, electrostatic attraction between two surfaces of elastomer films induces compressive strains, as shown in Figure 1. Since the elastomer maintains constant volume contraction in one direction will cause expansion in the other two. Most mechanisms use expansion perpendicular to the applied field because it will result in large displacements. The electrostrictive relaxor ferroelectric polymer actuators have high work density of 1 MJ/m^3 and strain up to 7–10%. These actuators generate high stress, around 45 MPa and frequency up to 100 Hz [27]. In electrostrictive relaxor ferroelectric actuators, the application of an electric field aligns polarized domains within the material. When the applied field is removed, the permanent polarization remains. Ferroelectrics are characterized by a curie point, a temperature above which thermal energy disrupts the permanent polarization. Field-driven alignment of polar groups produces reversible conformational changes that are used for actuation. The application of a field perpendicular to the chains leads to a transition between the non-polar and polar forms. The result is a contraction in the direction of polarization and an expansion perpendicular to it.

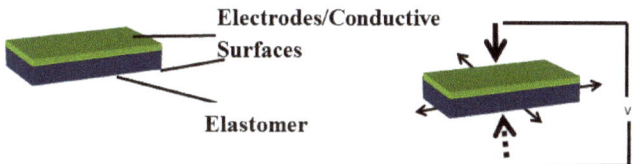

Figure 1. Schematic of dielectric elastomer mechanism with two electrodes: When a high electric field is applied to the electrodes the opposite charges attract squeezing the polymer into a different geometry causing an actuation of the device. "Reproduced with permission from [24], SPIE publications, 2000".

2.2. Ion Based Actuation

In these material's actuation is caused by the ion transport within the polymer material and exchange of ions between the actuator and an electrolyte solution. In common, the ionic EAPs need relatively low voltage for actuation (1–7 V) but the energies associated with these actuators are high because of the large amount of charge that needs to be transferred. Ion based actuators are most commonly fabricated with, conducting polymers (Conjugated polymers) and ionic polymer-metal composites (IPMC) [28].

Furthermore, IPMC contain an ion-exchange polymer film coated with metal electrodes. These metal electrodes are composed of platinum or silver nanoparticles. When the voltage is applied between two electrodes, the mobile cations move toward the oppositely charged electrode. This action results in swelling near the negative electrode, shrinkage near the positive electrode and bending of the actuator as can be seen in Figure 2.

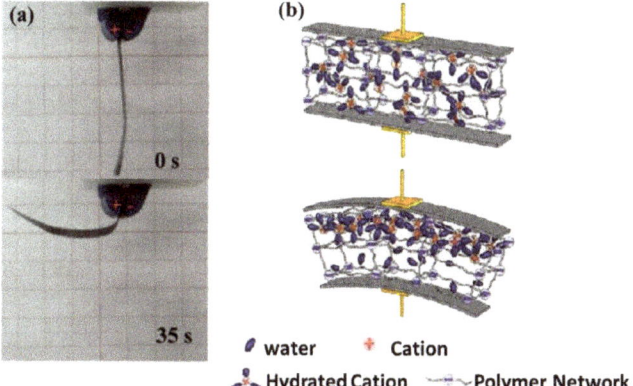

Figure 2. (a) The actuation of IPMC (b) the applied force cause the cation migration "Reproduced with permission from [29], Royal Society of Chemistry and Cambridge University Press [30]".

These actuators were reported with maximum actuation strains of 3.3% [26,31], and the stress of 30 MPa [25,32]. These actuators are actuated up to a frequency of 100 Hz [31]. The actuation mechanism of conducting polymers will be described in more detail in Section 3.1.1.

2.3. Pneumatic Actuation

The pneumatic artificial muscles (PAMs) are operated by air pressure and contract with inflation. These actuators consist of a soft membrane covered with a braided or fibrous filament structure. As the soft membrane is pressurized the volume is increased while expanding in the radial direction and contracting in the axial direction. The operating mechanism of PAMs can be described in two categories which are, (1) under a constant load and with varying pressure, and (2) with a constant gauge pressure and a varying load. As can be seen in Figure 3a the pressure is increased from P0 to P under constant weight of M which results in increasing the volume and decrease in length as demonstrated in Figure 3b. Actuation under the constant pressure is presented in Figure 3c,d. In this mode of operation weight is decreased from M to M0 under the constant pressure of P, which an actuator exhibits the maximum volume with the minimum length. The most widely used type of PAMs reported to date are the McKibben muscles [33,34]. These pneumatic actuators have high strength, high power-weight ratio, are economical and display high strength. However, the cycle life of these actuators is limited, due to the flexible membrane rupturing with stress. Pneumatic actuators have been reported with 25–30% actuation stroke and with actuation times of less than one second [35].

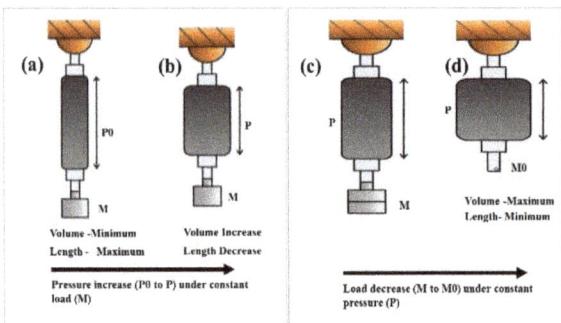

Figure 3. PAMs operation, (**a**) under constant weight (M) the pressure is increased to P, (**b**) volume is increased and length is decreased, (**c**) under constant pressure of P the weight is decreased to M0, (**d**) resulting in maximum volume and minimum length "Reproduced with permission from [36], Institute of Electrical and Electronics Engineers (IEEE), 2011".

2.4. Thermal Actuation

As the name suggests, thermal actuators are operated with the presence of heat. The first generation thermally actuated materials are shape memory alloys (SMAs), that "remember" their original shape and they returned to the original shape after being deformed and exposed to heat. The operating mechanism and fabrication details of SMAs are discussed in Section 3.3. Thermally actuated liquid crystal elastomers have the same working principles as of SMAs. In brief, phase changing and changing order alignment of liquid crystalline side chains generate stresses in the polymer backbone which result in actuation [25]. More importantly, liquid crystal elastomers display low stiffness. Therefore, a small change in the load can cause large displacements. In addition, actuation frequencies and loads on liquid crystal elastomers are limited by the tensile strength of these materials. The latest generation of thermally driven actuators is fabricated from synthetic polymer fibers with many outstanding properties. These actuators exceed natural muscle performance in many aspects and are recognized as one of the latest generations of artificial muscle actuators [17].

The actuation mechanism, fabrication and properties of these actuators are comprehensively described in Section 3.4.

2.5. Other Actuation Mechanisms

In addition to the more common actuator types listed above there are many other actuation mechanisms, such as electrochemical [16,37,38], electrostatic [39,40], optical [41], magnetic [42], hydraulic [43,44] and pH actuation [45,46].

3. Polymer Actuators in Smart Textiles

Some of the actuators described above consist of rigid components, robust operating systems and material properties which render them unsuitable for assembling into smart textiles. This section therefore will describe actuators with different mechanisms which have already been demonstrated in textiles mainly with polymer fiber actuators, such as conducting polymers [47] and shape memory polymers [48,49].

Helically arranged polymer actuators with amplified actuations have already been described in the literature. This encouraged researchers to consider these actuators in many further applications. The researchers employed an ancient technology of twisting which was able to produce highly twisted or coiled polymer fibers with giant actuations. The fiber types that have shown the capability to achieve these high actuation levels extended from twisted carbon nanotube (CNT) yarn to inexpensive commercially available fishing line and sewing threads [16,17,50,51]. This research was able to demonstrate reversible actuation cycles with high work capacity for the actuators. Therefore, actuating textile with helically arranged actuators can be further considered as an important approach for generating optimal force and strain. Hence, this material review is further intended to explore the properties of twisted and helically arranged actuator configurations, which has been successful with many materials, such as synthetic polymers and CNTs that have found potential applications in the area of smart textiles [14,18,38,52,53].

3.1. Conducting Polymers Actuators

The conducting polymers (CP) are also known as conjugated polymers, due to the altering single or double bonds in the polymer backbone. This is a class of electroactive polymers which are activated by ion transport [54]. CP actuators are normally actuated chemically or electrochemically and need electrolyte for their operation. Most of these semiconducting materials are doped with ions by chemical or electrochemical method.

3.1.1. Actuating Mechanism

The actuation mechanism of CP is very well described in many articles [25,55]. The CP actuators are operated under the mechanism of a dimensional change of the material which is caused by addition or removal of charge from the polymer structure.

The dimensional changes of these materials are achieved through the insertion of ions between polymers. The ion flux which is introduced by an electrolyte can cause swelling or contraction of the material as described below [25].

There are two major types of CP actuators classified as anionic and cationic driven. The CPs are produced by an oxidative polymerization process. During the chemical reaction, electrons are removed, and the monomers are put together by a chemical reaction to form the CP chains. Ionic cross links are formed with the polymer chains, due to insertion of anions (A-) which cause the material to be stiff and swollen, as shown in Figure 4a. Crosslinks formed by the bonding between anions and polarons (caused by the removal of electrons) enhance the inter-polymer bonding. The oxidized state of CP is reduced by applying a negative voltage either by way of Figure 4b or Figure 4d to the states Figure 4c or Figure 4e. When a small anion is used, the reduced state is achieved by way of Figure 4b as the anion is emitted causing the polymer to shrink as indicated in Figure 4c. With the oxidation the polymer is swollen from Figure 4c to Figure 4a through the process shown in Figure 4b. Thus, the mobile ions are anions in this mechanism, the actuators are named as "anion driven" actuators. The second mechanism takes place with the introduction of large anions during the fabrication of CP actuators. The immovable large anions are neutralized by inserting cations via process Figure 4d. This causes the polymer to further swell and achieve the status of Figure 4e. Due to the moving cations in this mechanism, these types of actuators are defined as "cation driven" actuators [55].

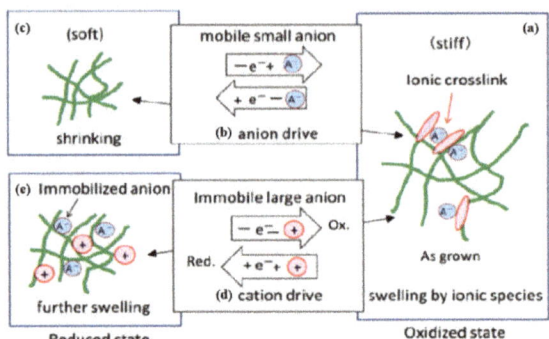

Figure 4. The mechanism of actuation in conducting polymers (**a**) oxidized state, (**b**) anion drive, (**d**) cation drive, (**c**) and (**e**) reduced state. "Reproduced from [55], Journal of physics, 2016".

3.1.2. Fabrication and Properties

CP actuators are typically fabricated through chemical or electrochemical polymerization of conducting materials. The common materials used for CP actuators are Polypyrrole (PPy), Polyaniline (PANi), and Poly (3, 4-ethylenedioxythiophene) (PEDOT)/poly styrene sulfonate (PSS). Due to the aromatic structure of these polymers which are shown in Figure 5, they are stable compared with other linear conducting polymers.

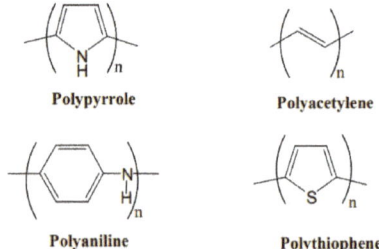

Figure 5. The chemical structure of conjugated CP in undoped form "Reproduced from [56], University of Wollongong Thesis Collection, 2009".

The materials that are used to fabricate these actuators have a strong influence over the actuator performance. PPy is the most popular material used for conducting polymer actuators. Predominantly, PPy is easily electrodeposited and it is feasible to obtain high conductive and tough films which provide high strain, force and long-life cycle [14,55,57]. Alternatively, PANi is prepared chemically by oxidative polymerization in bulk and the strain of actuators made from this material are lower when compared to PPy [58–61]. PEDOT:PSS is another material that has been used as a conductive coating in fabricating CP actuators. The fabrication of PEDOT:PSS actuators has been reported in combination with multi wall carbon nanotube, polyurethane/ionic liquid and Polyvinylidene fluoride [62–64]. CP actuators have been shown to exhibit both bending and linear movement. Linear actuators are fabricated by lamination of anionic and cationic driven actuators on a stretchable film. The fabrication of bi-layer and tri-layer conducting polymer actuators have also been reported in the literature [65–69]. The solvent and salts used in deposition and the electrolyte employed during actuation are the three major factors that play a significant role in determining the properties of these actuators. These actuators have a high tensile strength which can reach up to 100 MPa and with large stress up to 34 MPa [70]. Moreover, CP actuators are also able to withstand large stresses up to 34 MPa [71]. The strains of these actuators are typically 2–7% and the improvement for the CP actuators has been demonstrated even to reach up to 20% [72]. The strain rates of the CP actuators are low, since they are limited by the internal resistance of polymers, electrolytes and due to ionic diffusion rates [25]. Performance of CP actuators is weakened with the evaporation of the solvent during normal operation in air. As a resolution for evaporation, encapsulation methods were introduced to enhance the life time of these actuators [65,73]. Furthermore, actuators were introduced with internal ion conduction between active polymer layers instead of the external liquid electrolyte as an improvement. This research was demonstrated with PEDOT that shows the only deformation on actuation as can be seen in Figure 6 [74]. Consequently, CPs operated without an external electrolyte may increase their potential for incorporation into practical applications.

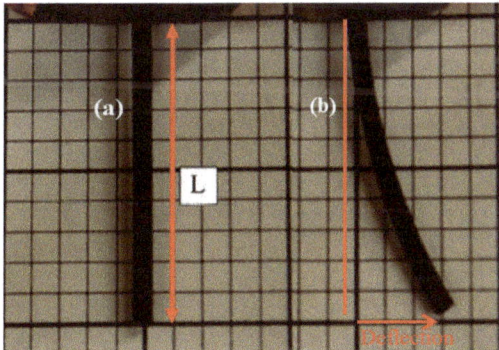

Figure 6. Actuator fabricated with PEDOT to provide deformation, (**a**) before and (**b**) after the application of 2V. The 20 mm length (L) actuator showed 6.5 mm deflection in open air "Reproduced with permission from [74], Elsevier, 2016".

Nevertheless, most of the linear CP actuators reported to date need encapsulation for an electrolyte which is an operational barrier [28]. The efficiency of these actuators is described to be low and their operational stability can be affected by the environmental conditions.

3.1.3. Conducting Polymer Based Actuating Textiles

The commercial availability of conducting polymer coated yarns makes them a practical option for use in actuating textiles [47]. A conducting polymer based actuating textile with different textile structures is presented in Figure 7. In this research a chemically synthesized PEDOT layer was deposited on the yarn/fabric as a "seed layer" to form a highly electrically conductive surface, followed by the deposition of the actuating PPy layer. This research verified the force amplification of actuators assembled into a woven textile structure and the increased strain by using a knitted textile. The research further confirms the mechanical stability of the CP actuators in textile structures [47]. This further outlines the different possibilities of a future improvement to the CP based actuating textile with enhanced features, such as conductivity and anisotropic movements.

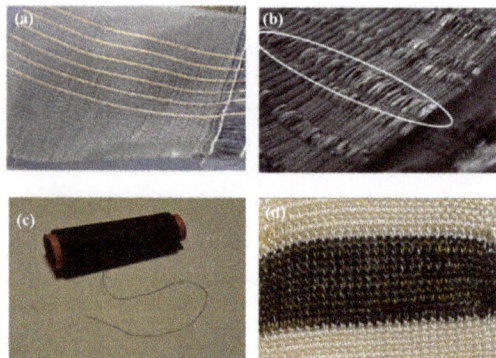

Figure 7. Processing and integration of electroactive textiles, (**a**) Copper monofilaments in weave fabric, (**b**) example of a custom weave with spacing (marked) that enables movements of yarns within the marked area, (**c**) a bobbin with industrially manufactured PEDOT-coated yarn, (**d**) a knitwear structure for respiratory monitoring with CP-coated yarns (black yarn) knitted together with normal (white) yarn "Reproduced from [47], Science Advances, 2017".

3.2. Carbon Nanotube Actuators

Research into Carbon Nanotubes (CNTs) over the last decade has demonstrated that CNTs have the capability to act as an actuating material powered electrochemically, electro thermally, electrostatically and/or optically [16,51,75–77]. The performance of CNT actuators has been increased with the research progress to improve the mechanical properties of CNT sheets and yarns. The following sections cover highlights in CNT actuator research.

3.2.1. Actuating Mechanism

The actuation of CNTs is achieved by mobile ions of a solvent within a polymer. An applied electric field leads to swelling or contraction of the CNT when these ions enter or leave the regions of the polymer. This is accomplished by dipping CNT in an electrolyte and applying a voltage (1–7 V) between the nanotubes. As the CNTs are electronically conductive, the ions are gathered onto the surfaces of the CNTs balancing the electronic charge as the potential has changed. This results in reformation of the electronic structure of the CNT which leads to dimensional changes.

The electrostatic actuation of CNTs is achieved by introducing a high level of charge injection. Electrostatic forces are generated, due to the interaction between the charges introduced into the CNTs instead of two electrodes as for electric field actuation [25]. The actuation of electrochemically powered CNT yarn has been demonstrated with the presence of electrolyte in several publications [16,38]. The actuation mechanism of CNT actuators was extensively studied and explained in the literature with twisted torsional artificial muscles reported by Foroughi et al. [16]. Moreover, CNT actuators with large torsional actuation at a high rotation rate were also demonstrated in this study. The large a scale actuation is achieved by applying a voltage between a counter electrode and a twisted multi wall carbon nanotube (MWNT) in an electrolyte. The contraction of the reported CNT is due to the volume expansion caused by ion insertion which provides a 1% lengthwise contraction with respect to the initial length. A scanning electron microscopic (SEM) image of the twisted MWNT symmetrically twist-spun from an MWNT forest is shown in Figure 8a. The actuation mechanism in brief can be described as a partial untwist of the yarn during the charge injection which is changing the geometrical configuration of the yarn from Figure 8b1 to Figure 8b2. This is associated with the yarn volume expansion after the large positive or negative charge insertion which results in a lengthwise contraction. This research provides further evidence for twist-spun nanotube yarns driven by internal pressure, due to ion insertion [16].

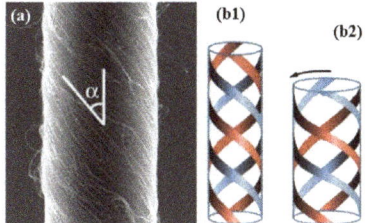

Figure 8. (a) SEM of twisted twisted carbon nanotube (CNT) yarn, (b) Schematic of the yarn volume expansion during the charge injection "Reproduced with permission from American Association for the Advancement of Science [16], 2011".

Meanwhile, electrothermally driven CNT actuators were reported in the literature overcoming the necessity for the presence of electrolyte for actuation. The electrothermal actuation of CNT was achieved through combining with other polymers which have the ability to thermally expand and contract, such as phase change materials like paraffin wax [78] or with CNT network in silicone polymer elastomer [79]. In general, the electrothermal actuation mechanism of hybrid yarn is driven by volume expansion of the guest polymer materials which are merged with the CNT. Nevertheless, electro thermally driven hybrid CNT actuators need comparatively high applied voltage compared to electrochemically driven actuators [76].

3.2.2. Fabrication and Properties

The electrochemical actuation of CNT was first demonstrated by Baughman et al. for CNT sheets [37]. The research was validated with single-walled nanotube (SWNT) sheets which generated higher stresses and strain than natural muscles. This study opened up possible new dimensions in actuator technology. Thereafter, CNT actuators with un-oriented CNT sheets were demonstrated by a group of researchers. These actuators with low modulus and strength generated around 0.2% stroke and stress 100 times more than skeletal muscle. This study further demonstrated electrostatically driven actuators with 220% stroke [51]. The above research demonstrated actuation for CNT in form of sheets. Meanwhile, a process for the continuous production of CNT yarn fabrication was introduced. The fabrication of CNT yarn evolved by combining the ancient technology of twist insertion during the spinning process. As can be seen in Figure 9a, the CNT yarn is drawn from a vertically aligned MWNT forest. Then the CNT yarn is twisted by a spinning machine as presented in Figure 9b. The schematic Figure 9c shows the magnified view of yarn drawing, twisting and winding during the spinning process. The SEM image in Figure 9d shows the CNT yarn was drawn and twisted simultaneously during the fabrication process. This procedure was able to produce a high strength, multi plied torque stabilized CNT yarn in which the strengths are greater than 460 MPa [50]. Further, the twisted MWNT actuators were demonstrated with high torsional actuation per muscle length with high rotation rates which provided a breakthrough for many types of helically arranged actuators. The twisted CNT actuator was mainly demonstrated for torsional actuation that demonstrated a practical application for a prototype mixer [16].

As mentioned above, the torsional or the tensile actuation of CNTs are achieved as a result of a volume change of the yarn. To accommodate the volume changes, an electrolyte or a guest material should be introduced into the CNT yarn structure. In contrast, the electrolyte used in electrochemically driven actuators adds more volume to the actuator system. Therefore, rather than fabricating these actuators using a sole material, researchers had shown an interest to fabricate CNT hybrid actuators in solid states. As a result, identical anode and cathode yarns were fabricated by permeating the electrolyte and electronically insulating the surface of the yarn to prevent any electrical shorting. The microscopic images of the CNT solid state actuators are presented in Figure 10. As shown in the figure, all solid state actuators were fabricated by plying anode and cathode yarns together [76].

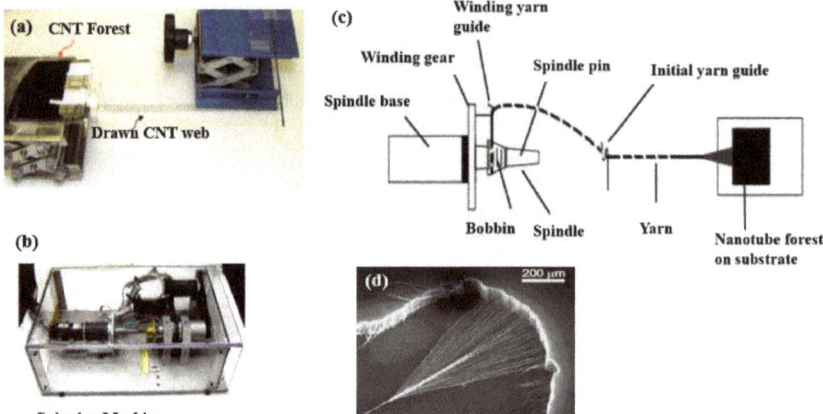

Figure 9. (a) The CNT drawn from CNT forest, (b) spinning machine, (c) the schematic diagram of the CNT spinning "Reproduced from [56], University of Wollongong Thesis Collection, 2009". (d) SEM image of CNT yarn being drawn and twisted "Reproduced with permission from [50], American Association for the Advancement of Science 2004".

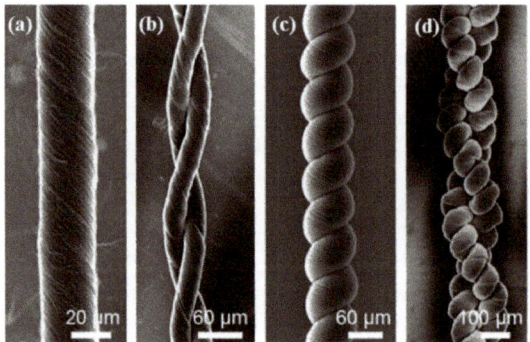

Figure 10. Twisted multi wall carbon nanotube (MWNT) yarn structures, (a) Scanning electron microscopic images of single yarn, (b) two ply yarns, and (c) Single coiled yarn and (d) plied coiled yarns "Reproduced with permission from [76]), American Chemical Society2014".

3.2.3. CNT Based Actuating Textile

MWNT yarns largely retain the twist when yarn ends are released compared to conventional textile yarns. Studies have found that these yarns can retain their twist up to the breaking point [50]. Accordingly, highly twisted yarns were demonstrated for plying, knitting and knotting, as well as shown in Figure 11 [50].

Moreover, electrochemically driven plied actuators were reported by Lee et al. [75]. These actuators provided a tensile contraction of 11.6% and 5% for parallel and braided muscles respectively, which were driven electrochemically without a liquid electrolyte. This research further progressed to produce an energy conserving actuator with 16.5% contraction which is the highest reported to date. Theses actuators eliminate the electrolyte bath by replacing it with an ionically conducting gel, as shown in Figure 12b. The gel insulates the anode and cathode yarns while providing ionic conduction [75].

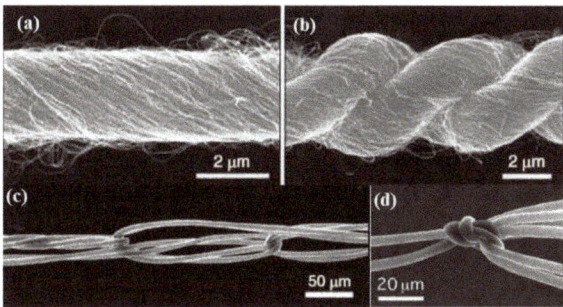

Figure 11. SEM images of (**a**) single twisted, (**b**) two-ply, (**c**) knitted and (**d**) knotted MWNT yarns "Reproduced with permission from [50], American Association for the Advancement of Science, 2004".

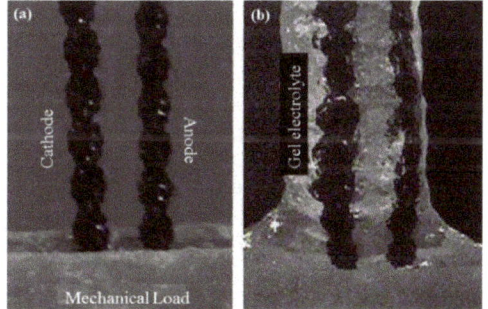

Figure 12. Optical microscopic images of parallel arranged actuators (**a**) before, (**b**) after coating with gel electrolyte to accommodate ion conduction" Reproduced with permission from [75], WILEY-VCH, 2017".

Even though these studies demonstrated technical feasibility, the cost of CNT yarns can be the major drawback in the production of a CNT based actuating textile.

3.3. Shape Memory Alloy (SMA) Actuators

Thermally actuated shape memory alloys (SMAs) are a class of materials that can "remember" their original shape. SMA actuators with both linear or rotary motions are reported in the literature that provided a great impact for thermally driven actuator technology [80].

3.3.1. Actuating Mechanism

The operating mechanism of SMA actuators has not been fully verified, since direct observation of their dynamic behavior in a wide range of temperature is difficult. The actuation of SMA occurs due to a change in the atomic structure between two phases: The low temperature (martensite) and high temperature (austenite), as shown in Figure 13. The actuating mechanism of SMA is achieved by training the material to remember a definite shape at high temperature. Both phases are identical in chemical composition, but when the material is deformed at low temperature the residual strain can be recovered by heating it to the austenite state. This type of SMAs can only remember the parent high temperature phase, and so are referred to as SMAs with one-way shape memory effect. The actuators with two way shape memory effect can perform in two stable phases, i.e., both in high temperature and low temperature [25]. Two way SMAs can provide tensile force much lower than the contraction force and the strain exhibited is half of that can be seen in one way type [81].

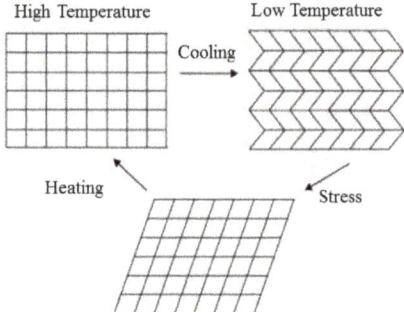

Figure 13. Grid-like representation of shape memory alloys (SMA) structure "Reproduced from [82], InTech., 2010".

3.3.2. Fabrication and Properties

A limited number of raw materials were used to fabricate SMA actuators in the literature. The Nitinol (Ni-Ti) is the most widely used SMA although Copper and iron based SMAs are also employed in some applications. The material selections for SMAs are highly dependent on their transformation temperature. Relatively, Ni-Ti is expensive and copper alloys are less costly but not as widely used, due to the lower fatigue tolerance and thermomechanical instability [83]. The attractive properties of SMA actuators, such as low operating voltage, clean, silent and having a long actuation cycle life have enabled them to be used in many applications [82]. SMAs exhibit a high energy (work) density which is around 1000 KJ/m^3. These actuators operate at very high strain rates (around 300% per second) responsive and exhibit large deformations (around 5%) [26]. Furthermore, SMAs are very responsive and can deliver large strokes. The operating frequencies of these actuators are dependent upon the rate of cooling and heating of SMA to promote phase change. Conversely, exhibiting energy loss during phase transformation can cause a hysteretic behavior to the SMA including nonlinear actuation, parameter uncertainties and their relative costs restrict their use in commercial applications [84].

3.3.3. SMA based Actuating Textiles

An SMA based actuating textile designed for self-recovery by weaving and knitting textile structures with embedded Ni-Ti wires was introduced by Carosio et al. [48]. The fabricated woven textile structure was shown to display self-ironing with the presence of Ni-Ti wires. The fabric was crushed, as shown in Figure 14a, and then was able to exhibit a self-shape recovery as can be seen in Figure 14b. This research further highlights the successful combination of Ni-Ti wires in a woven fabric structure [48].

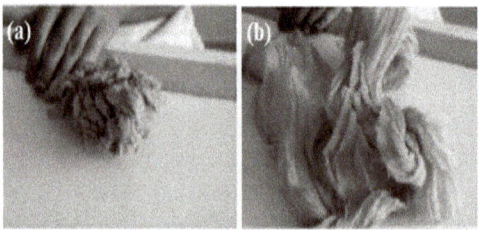

Figure 14. The Ni-Ti embedded fabric (a) crushed and (b) self-recovered "Reproduced with permission from [48], IOS Press, 2004".

Further, an analytical model using SMA in a garter knit structure was presented by Juliana et al. [51]. A prototype knit textile was fabricated and tested within the range of forces as a characterisation of the textile. The knitted textile fabricated from Flexinol actuators was able to achieve larger strains (around 51%) at moderate forces and usable strains (around 4.1%) at the enhanced force of 12 N, compared to the single actuator alone with 4% strain at 5.8 N [49].

3.4. Twisted and Coiled Synthetic Fibre Actuators

Synthetic fibers are designated as "man-made fibers". These are popular in many practical applications, due to interesting properties, such as high tensile strength, high modulus and shear stability [85]. The precursor fibers used to fabricate coiled actuators are readily being used in high strength applications, such as fishing, apparel and sewing. The high degree of polymer alignment of these fibers provides them with high strength. Moreover, forming these fibers in a twisted fashion and arranging the polymer chains helically provides for a thermally persuaded length change during untwisting. The phenomenon for actuation of these materials will further be described in the section below.

3.4.1. Actuating Mechanism

Synthetic fibers are produced from a process called "polymerization" followed by fiber drawing. Upon drawing, the crystalline blocks of the polymer become increasingly aligned along the draw direction. The drawn polymers will consist of an amorphous region, tie molecules and inter crystalline bridges, as shown in Figure 15. The amorphous region contains floating chains and polymer chains which are attached to the crystalline region at one end and loops, which starts and end at the same crystalline region. The tie molecules joining one crystalline block to another block increases with both number and steadiness by increasing draw ratio. The crystalline regions of polymer fibers have a small degree of negative thermal expansion. Fiber direction aligned polymer chains in non-crystalline regions are less constrained and thus they can cause larger reversible contractions when heated. This reversible contraction is amplified by inserting twists and coiling the yarns.

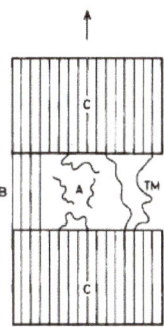

Figure 15. Schematic diagram showing the structure of a highly oriented semi crystalline polymer. (C) crystalline region; (B) bridges; (A) amorphous region; (TM) tie-molecules "Reproduced with permission from [86], John Wiley and Sons, 1981".

The giant actuation of these actuators is achieved through partial untwisting of the twisted fibers [17]. The untwisting of twisted fibers provides an expansion in the radial direction which leads to a contraction in the fiber axis direction.

3.4.2. Fabrication and Properties

High strength polymer fibers, such as nylon, polyester and polyethylene, are anisotropic materials and considered as raw materials for these actuators. The fabrication procedure of these actuators was fully described in research work by Carter S. Haines et al. [17]. The precursor fibers (Figure 16a) were twisted until they get coiled, as shown in Figure 16b or they can be fabricated by wrapping the twisted fiber around a mandrel as can be seen in Figure 16e. The actuator structure was set using an annealing procedure to retain the helical shape. Furthermore, actuators can be tailor made to achieve the desired actuation based on fundamental studies.

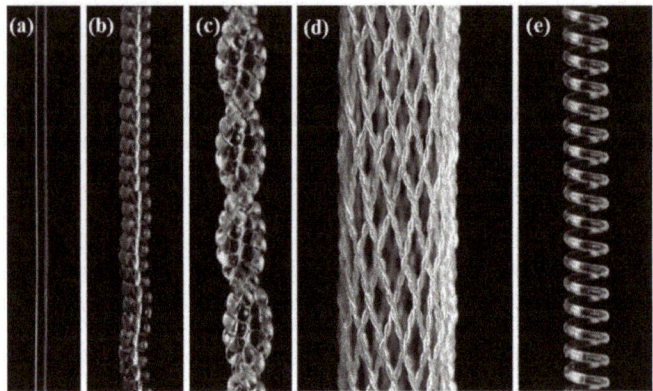

Figure 16. The actuators (**a**) a non-twisted monofilament, (**b**) after coiling the monofilament, (**c**) a two –ply muscle formed from the coil, (**d**) a braid formed from 2-ply muscles, (**e**) a coil formed by inserting a twist "Reproduced with permission from [17], American Association for the Advancement of Science, 2014".

Figure 17 shows the bulk-produced coiled actuators manufactured by a continuous process where (a1) is a spool of the non-conductive actuator and (a2) is a spool of the conductive actuator. The conductive actuator is fabricated by wrapping with insulated copper wire for electrothermal heating. The continuous production possibility of these actuators will further enhance the feasibility of fabricating them into textiles.

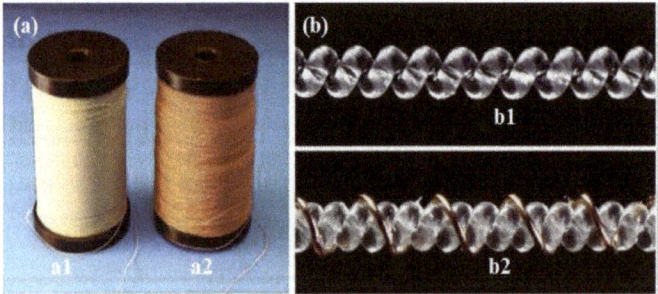

Figure 17. Coiled polymer actuators produced by a continuous process, (**a1**) spool of non-conductive actuator and the optical image of non-conductive actuator is shown in (**b1**), (**a2**) spool of conductive actuators produced by wrapping with an insulated copper wire, as shown in optical image (**b2**) "Reproduced from [19], Proceedings of the National Academy of Sciences, 2016".

These coiled synthetic polymer actuators exhibited a 49% maximum lengthwise contraction. Furthermore, these actuators were able to lift loads over 100 times heavier than a human muscle of the same length and weight. In addition, they can generate 5.3 kW/kg of mechanical work, (similar to that produced by a jet engine) with the highest operating frequency of 7.5 Hz reported to date. The low cost, less-hysteretic behavior, ease of handling, high tensile strength and other exhibited performance characteristics are some additional favorable properties of these actuators [17]. Further research of synthetic polymer actuators was published by Cater S. Haines et al., which discussed the practical opportunities and challenges of artificial muscles. This research highlights the limiting factors of the tensile actuation and the further improved spiral shape actuator which was fabricated with 200% tensile actuation [19]. Thus, the coiled actuators have been widely investigated by researchers for textile fabrication.

3.4.3. Twisted Polymer based Actuating Textiles

Twisting and coil formation of polymers offer high-performance actuators which provide promising materials in designing a high-performance actuating textile. A model textile has been demonstrated for the first time in the literature from the twisted actuators with nylon fishing line, as shown in Figure 18 [17]. The textile was weaved from silver-plated nylon for electrothermal heating (brown in color) and polyester, cotton yarns (white and yellow in color) in the weft direction and nylon coiled actuators were used as the warp yarn.

Figure 18. An actuating textile woven from conventional polyester, cotton, and silver-plated nylon yarn (to drive electrothermal actuation) in the weft direction "Reproduced with permission from [17], American Association for the Advancement of Science, 2014".

The textile actuation was achieved via heating the textile electrically which provides a gateway for fabricating novel actuating textiles. Thereafter, actuating textiles were formed using traditional textile fabrication methods with the recent research of Hanes et al., as shown in Figure 19 [19].

This research successfully combined the actuators in woven, stitched and knitted textile structures. These textiles were fabricated with non-electrically conductive actuators. The researchers have recommended these textiles in applications, such as porosity changing textiles and breathable curtains.

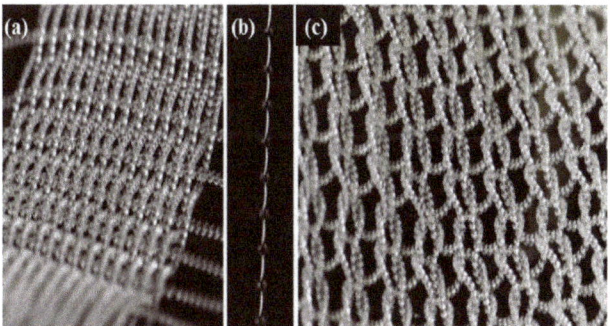

Figure 19. (a) Woven fabric made from coiled, 225-μm-diameter nylon sewing thread coils, (b) stitches made by sewing the coiled fiber in into a polymer sheet using a conventional sewing machine, (c) machine-knitted textile made from a coiled 225-μm-diameter nylon sewing thread. "Reproduced from [19], Proceedings of the National Academy of Sciences, 2016".

Furthermore, nylon actuators were recently demonstrated in a bionic bra developed to minimize breast discomfort during exercise [87]. The woven actuators were used as active materials to control the breast movement, as shown in Figure 20.

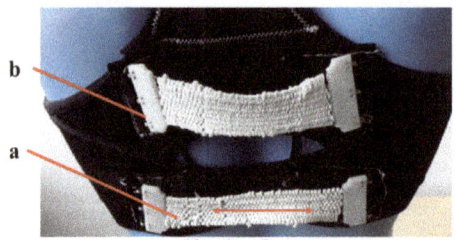

Figure 20. The bionic bra fabricated with woven actuators (**a**) the actuator placement in the bra woven with an electrically conductive yarn for heating (**b**) 3D printed actuator connector.

The actuators were heated by weaving them with a conductive yarn. A single actuating fiber was able to generate around 0.6 N force following heating to 75 °C and a woven textile actuator with the nine, parallel actuating fibers was able to generate around 3 N force heating to the same temperature.

3.5. Knitted CNT/Spandex Yarn as Smart Textiles

More interestingly, an electrothermally activated "clever yarn" was invented by overcoming the technical obstacles by Foroughi et al., as shown in Figure 21 [13]. A highly stretchable, actuatable textile was produced by wrapping spandex filaments (SPX) with CNT yarns to give the actuating performance and conductivity respectively. This knitted textile structure exhibits 33% contraction and mechanical work output of 1.28 kW/kg which exceeds that of skeletal muscle. This research presents adjusted electrical conductivities by changing the SPX/CNT ratio and hysteresis free resistance was obtained by changing the tensile strain. A hybrid SPX/CNT based actuating textile opened a new dimension into manufacturing actuating textiles using an existing textile fabrication method. Further, this was recommended for applications where it was required to apply force or pressure to the wearer [13]. The actuating textile was heated by applying a voltage of 12 V and current of 0.25 A. Further, this research demonstrated the feasibility of using coiled synthetic fiber actuators in smart textiles.

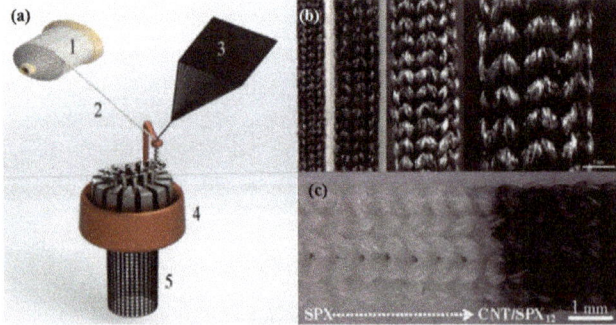

Figure 21. (**a**) Schematic of the process for producing a knitted CNT/SPX (wrapping spandex filaments) textile. The illustrated items are (1) a spool of SPX fibers, (2) an n-fiber SPX yarn, (3) a CNT forest, (4) a circular knitting machine, and (5) a knitted CNT/SPX textile. A CNT ribbon drawn from a CNT forest was wrapped around SPX fibers and knitted in the knitting machine to produce the circular knitted textile, shown in (**b**) and (**c**). "Reproduced with permission from [13], American Chemical Society, 2016".

4. Overview in Different Actuation Mechanism for Smart Textiles

The actuating mechanism should be selected with a major emphasis on end user requirement. This section is focused to discuss the suitability of different actuation mechanisms for an actuating textile for high tech applications including biomedical, soft robotics and apparel. Herein, we are appraising above described popular actuation mechanisms; electric field, ion based, pneumatic and thermal means for a workable textile fabrication.

Electric field actuation is caused by electrostatic attraction. Therefore, it requires two surfaces or alignment of polarized domains which need voltages as high as 1 kV. Generally, there is need of an amplifier to convert line or battery voltages up to kV potentials, which adds cost and consumes volume. Thus, the cost, size and safety measures may prohibit electric field actuators for applications in small portable (e.g., handheld) devices. All these limitations can be a concern in smart textiles, as well as in bio medical and toy applications [25,26].

Ion based actuation requires electrolyte to be presented in the polymer structure. Therefore, assembling them in a smart textile would need a configuration to retain the electrolyte medium. Actuator arrangements described in the literature without the liquid electrolyte need sophisticated manufacturing procedures and actuation mechanisms which add more cost and operating barriers to the system. Moreover, low efficiencies are one of the key disadvantages of these actuator types. The main disadvantages of pneumatic actuators, such as McKibben, are that, they need a compressor or pump and their dynamic behavior is nonlinear. Consequently, they are difficult to make into a textile, and a robust control mechanism is needed to achieve the desired motion [88,89]. Therefore, the feasibility of incorporating actuators with these major actuating mechanisms into smart textile will present many challenges.

In contrast, most of the research has focused on the use of thermally driven actuators in smart textiles [13,17,19,48,49,87]. This may be mainly due to the utilization of electrothermal heating as a reliable and clean source of energy. Many studies have been reported for textiles combined with different types of electrical conductors for smart textiles and electrothermal heating applications [90–95]. Furthermore, the outstanding properties of thermally driven actuators make them an attractive prospect in high end future applications. Thermally driven synthetic coiled actuators show 49% contraction which will enable them to be used in a highly contractible actuating textile. This high contractibility is able to generate a high pressure which makes it an attractive proposition in many applications. Since, it will produce a textile structure which is generating high work output per unit area. The cost of raw materials and cost of processing sets the final price limit for the textile. The use of inexpensive synthetic polymers will be advantages for producing an actuating textile at low cost which will increase affordability and market demand. Moreover, the durability and demonstrated operating consistency will increase the feasibility for their use in applications, such as bio- medical where reliability is paramount. The high tensile strength and the reversible actuation over one million cycles enables the production of a durable textile with less damage to the actuator system during operation. Additionally, the low hysteresis behavior of synthetic polymer coils increases the possibility of producing an easily controllable textile which will exhibit a consistent actuation in heating and cooling cycles. These outstanding properties point to exciting prospects for the use of coiled synthetic actuators in future high-performance actuating textiles. Table 1 provides a summary of the properties of different actuating mechanisms, actuators, their advantages and drawbacks.

Table 1. The summary of different actuators with their actuating mechanisms, actuator properties, advantages and disadvantages.

Actuating Mechanism	Actuator Type	Strain	Stress (Mpa)	Work Capacity	Advantages	Disadvantages	References
Electric Field Actuation	Dielectric elastomeric actuators	10–380%	7.7	150 kJ/m^3	Simple in mechanism and construction	High voltages	-
					Large strains	Cost and consumes volume	-
					High efficiencies (30%)	-	[22,23]
					High bandwidths	-	
					Low current		
					Low cost		
	Electrostrictive polymers	7%	45	320 kJ/m^3	High work density	High voltages cycle life is unclear	[26,27]
					High Stress		
Ion based Actuation	Conducting polymer	12%	34	100 kJ/m^3	Low Voltage	Need encapsulation	[25,70,72]
					High Stress	Low Efficiencies	-
					High work density	-	
	Ionic Polymer Metal Composites	3.30%	30	5.5 kJ/m^3	Low Voltage	Need encapsulation	[25,26,31,32]
Pneumatic Actuation	Mckibben	25–30%	-	-	Fast	Difficulty to control	-
					High strain	Bulky operating method	[35]
Thermal Actuation	Shape Memory Alloys	8%	200	1000 kJ/m^3	High stress Low Voltage High work density	Difficult to control Large currents Low efficiencies	[26,84]
	Twisted Synthetic Fibers	49%	-	2.48 kJ/kg	High Strain		
					High work output per kg Inexpensive	Limited Operating temperatures (250 °C)	[17]
					Light weight	-	
					Flexible	-	
Electrochemical	Carbon Nanotubes Yarns	16%	2	2 kJ/m^3	High Stress	Expensive	-
					Low Voltage	-	[26,75]
					High temperature range	-	

5. Future Outlook

An electrically operated high-performance actuating textile with high strain and force will benefit many future applications. However, the fabrication of conductive textiles is limited by many technical and nontechnical parameters and it is important to analyze the contests associated with fabrication. The processes of weaving or knitting electrically conductive actuators for electrothermal heating are normally associated with several challenges, such as an increase of electrical resistance with tensile strain and changes in the conductive path ways of textile yarns or fabrics. In addition, the weaving of conductive yarns/wires involves high strains, process constraints, and long-term stability of intersection points are some of the other limitations of bulk manufacturing of actuating textiles. Electrical shorting between two yarns at their mechanical intersection is another key limitation in establishing electrical links [96]. Damaging of electrical connections during washing, and other activities performed by the user are some other challenges in smart textile applications. As a result, some researchers have investigated the used of surface modified conductive textile yarns for their hydrophobicity and electrical encapsulation. As examples, the textile cables can be coated with silicone substrates with improved mechanical properties which will minimize the conductivity and electrical shorting limitations described above [97]. Further, a method of encapsulation for washable, reliable and wearable electronics was demonstrated by Tao et al. which was focused on two types of silicone where the devices were able to perform after washing [98]. Moreover, the actuation frequency can be lowered with cooling in normal air. Therefore, incorporating cooling material is important to maintain a consistent frequency during operation [99]. Heating might be another limiting factor as the human body can only tolerate a certain temperature range. This temperature range will differ with the application type and the area of the body exposed to the textile. Hence, the limiting factors of electro thermally driven textiles need to be well controlled in order to produce a high-performance actuating textile.

6. Conclusions

Development of materials for the preparation of actuating materials is an important enabling step towards their application, particularly in smart textiles. We have summarized the history of the emergence of actuating materials, categories, and preparation and fabrication methods for the recent development of smart textiles, as well as their current/future of applications. Smart textiles based on different actuating mechanisms and a comprehensive study on polymer actuators has been reviewed. The compatibility of diverse actuating mechanisms in actuating textile was showed that the thermally driven actuators can be considered as a potential actuator type to incorporate into actuating textiles. Moreover, thermally driven twisted synthetic fiber actuators with high actuation strain and work capacity will provide extraordinary features to a high-performance actuating textile. Thermally operating actuators are fabricated based on man-made fibers, such as nylon, polyester, and spandex which can be manufactured using conventional textile processing. In addition, the thermal energy for actuation can be harvested from electric power as a reliable method of operation. The electrical Joule heating method for textile can be achieved by incorporating conductive materials to the textile structure. The processes and materials described above need to be evaluated when considering the fabrication of electrically operated actuating textiles. The introduction of guest materials to achieve desired application properties is also a key area to be considered during fabrication. Most importantly, thermally driven actuators should be evaluated for the most efficient and even means of heating. Actuators heated electrically by connecting with a conductive yarn or conductive coatings are some of the technically stable methods which have enormous potential. Furthermore, the possibility of using contractile polymers in an artificial heart has been investigated and there is a high possibility of employing an electrically operated actuating material structure in biomedical applications [100–102]. A well fabricated actuating textile will find multiple applications possibilities, such as biomedical, prosthetics, soft robotics, and smart apparel which can make a significant impact in many areas. Although recent development in smart textiles appears extremely promising, there still remain challenges to improve their properties and performance to become adequate for a practical and commercial application.

Author Contributions: Both authors have involved with the same contributions to this work.

Funding: The Australian Research Council under Discovery Early Career Researcher award, funding number: DE130100517, Javad Foroughi.

Acknowledgments: The authors would like to thank the Australian Research Council under Discovery Early Career Researcher award (J. Foroughi DE130100517). This research has been conducted with the support of the Australian Government Research Training Program Scholarship.

Conflicts of Interest: The authors declare no conflict of interest.

References

1. Syduzzaman, M.; Patwary, S.U.; Farhana, K.; Ahmed, S. Smart Textiles and Nano-Technology: A General Overview. *J. Text. Sci. Eng.* **2015**, *5*, 181. [CrossRef]
2. Van Langenhove, L.; Hertleer, C. Smart clothing: A new life. *Int. J. Cloth. Sci. Technol.* **2004**, *16*, 63–72. [CrossRef]
3. Foroughi, J.; Spinks, G.M.; Wallace, G.G. Conducting Polymer Fibers. In *Handbook of Smart Textiles*; Tao, X., Ed.; Springer: Singapore, 2015; pp. 31–62.
4. Park, S. Smart Textiles: Wearable Electronic Systems. *MRS Bull.* **2003**, *28*, 585–591. [CrossRef]
5. Sahin, O.; Kayacan, O.; Bulgun, E.Y. Smart textiles for soldier of the future. *Def. Sci. J.* **2005**, *55*, 195–205. [CrossRef]
6. Ramdayal; Balasubramanian, K. Advancement in Textile Technology for Defence Application. *Def. Sci. J.* **2013**, *63*, 331–339.
7. Krishnan, M.; Kannan, G. Polygon Shaped 3G Mobile Band Antennas for High Tech Military Uniforms. *Adv. Electromagn.* **2016**, *5*, 7–13. [CrossRef]
8. Zhang, X.; Tao, X. Smart textiles: Passive smart. *Text. Asia* **2001**, *32*, 45–49.
9. Zhang, X.; Tao, X. Smart textiles: Very smart. *Text. Asia* **2001**, 35–37.
10. Zhang, X.; Tao, X. Smart textiles: Active smart. *Text. Asia* **2001**, 49–52.
11. Stoppa, M.; Chiolerio, A. Wearable Electronics and Smart Textiles: A Critical Review. *Sensors* **2014**, *14*, 11957–11992. [CrossRef]
12. Stoychev, G.V.; Ionov, L. Actuating Fibers: Design and Applications. *ACS Appl. Mater. Interfaces* **2016**, *8*, 24281–24294. [CrossRef] [PubMed]
13. Foroughi, J.; Spinks, G.M.; Aziz, S.; Mirabedini, A.; Jeiranikhameneh, A.; Wallace, G.G.; Kozlov, M.E.; Baughman, R.H. Knitted Carbon-Nanotube-Sheath/Spandex-Core Elastomeric Yarns for Artificial Muscles and Strain Sensing. *ACS Nano* **2016**, *10*, 9129–9135. [CrossRef] [PubMed]
14. Foroughi, J.; Spinks, G.M.; Wallace, G.G. High strain electromechanical actuators based on electrodeposited polypyrrole doped with di-(2-ethylhexyl)sulfosuccinate. *Sens. Actuators B Chem.* **2011**, *155*, 278–284. [CrossRef]
15. Foroughi, J.; Spinks, G.M.; Wallace, G.G. A reactive wet spinning approach to polypyrrole fibres. *J. Mater. Chem.* **2011**, *21*, 6421–6426. [CrossRef]
16. Foroughi, J.; Spinks, G.M.; Wallace, G.G.; Oh, J.; Kozlov, M.E.; Fang, S.; Mirfakhrai, T.; Madden, J.D.W.; Shin, M.K.; Kim, S.J.; et al. Torsional Carbon Nanotube Artificial Muscles. *Science* **2011**, *334*, 494–497. [CrossRef] [PubMed]
17. Haines, C.S.; Lima, M.D.; Li, N.; Spinks, G.M.; Foroughi, J.; Madden, J.D.W.; Kim, S.H.; Fang, S.; Jung de Andrade, M.; Göktepe, F.; et al. Artificial Muscles from Fishing Line and Sewing Thread. *Science* **2014**, *343*, 868–872. [CrossRef] [PubMed]
18. Lima, M.D.; Li, N.; de Andrade, M.J.; Fang, S.; Oh, J.; Spinks, G.M.; Kozlov, M.E.; Haines, C.S.; Suh, D.; Foroughi, J.; et al. Electrically, Chemically, and Photonically Powered Torsional and Tensile Actuation of Hybrid Carbon Nanotube Yarn Muscles. *Science* **2012**, *338*, 928–932. [CrossRef] [PubMed]
19. Haines, C.S. New twist on artificial muscles. *Proc. Natl. Acad. Sci. USA* **2016**, *113*, 11709–11716. [CrossRef] [PubMed]
20. Tondu, B. Artificial muscles for humanoid robots. In *Humanoid Robots, Human-Like Machines*; InTech: London, UK, 2007.
21. Huber, J.E.; Fleck, N.A.; Ashby, M.F. The selection of mechanical actuators based on performance indices. *Proc. R. Soc. Lond. A: Math. Phys. Eng. Sci.* **1997**, *453*, 2185–2205. [CrossRef]
22. Bar-Cohen, Y. Current and future developments in artificial muscles using electroactive polymers. *Expert Rev. Med. Devices* **2005**, *2*, 731–740. [CrossRef]
23. Vincenzini, P.; Bar-Cohen, Y.; Carpi, F.; Vincenzini, P. *Actuators Using Electroactive Polymers: Actuators Using Electroactive Polymers: Cimtec 2008*; Trans Tech Publications, Limited: Durnten, Switzerland, 2008.
24. Kornbluh, R.D.; Pelrine, R.; Pei, Q.; Oh, S.; Joseph, J. Ultrahigh strain response of field-actuated elastomeric polymers. In Proceedings of the SPIE's 7th Annual International Symposium on Smart Structures and Materials, Beach, CA, USA, 6–9 March 2000; p. 14.

25. Mirfakhrai, T.; Madden, J.D.W.; Baughman, R.H. Polymer artificial muscles. *Mater. Today* **2007**, *10*, 30–38. [CrossRef]
26. Madden, J.D.; Vandesteeg, N.A.; Anquetil, P.A.; Madden, P.G.; Takshi, A.; Pytel, R.Z.; Lafontaine, S.R.; Wieringa, P.A.; Hunter, I.W. Technology: Physical principles and naval prospects. *IEEE J. Ocean. Eng.* **2004**, *29*, 706–728. [CrossRef]
27. Wallmersperger, T.; Kröplin, B.; Gülch, R. *Electroactive Polymer (EAP) Actuators as Artificial Muscles-Reality, Potential, and Challenges*; Modelling and Analysis of Chemistry and Electromechanics; Spie Press: Bellingham, WA, USA, 2004; Volume PM 136.
28. Kim, K.J. *Biomimetic Robotic Artificial Muscles. [Electronic Resource]*; World Scientific: Singapore; Hackensack, NJ, USA, 2013.
29. Tang, Y.; Xue, Z.; Xie, X.; Zhou, X. Ionic polymer–metal composite actuator based on sulfonated poly (ether ether ketone) with different degrees of sulfonation. *Sens. Actuators A Phys.* **2016**, *238*, 167–176. [CrossRef]
30. Park, I.-S.; Jung, K.; Kim, D.; Kim, S.-M.; Kim, K.J. Physical principles of ionic polymer–metal composites as electroactive actuators and sensors. *MRS Bull.* **2008**, *33*, 190–195. [CrossRef]
31. Nemat-Nasser, S.; Wu, Y. Comparative experimental study of ionic polymer–metal composites with different backbone ionomers and in various cation forms. *J. Appl. Phys.* **2003**, *93*, 5255–5267. [CrossRef]
32. Shahinpoor, M.; Kim, K.J. Ionic polymer-metal composites: I. Fundamentals. *Smart Mater. Struct.* **2001**, *10*, 819–833. [CrossRef]
33. Laksanacharoen, S. Artificial Muscle Construction Using Natural Rubber Latex in Thailand. In Proceedings of the 3rd Thailand and Material Science and Technology Conference, Bangkok, Thailand, 10–11 August 2004; pp. 1–3.
34. Agerholm, M. The "artificial muscle" of mckibben. *Lancet* **1961**, *277*, 660–661. [CrossRef]
35. Sangian, D. New Types of McKibben Artificial Muscles. Ph.D. Thesis, School of Mechanical, Materials and Mechatronic Engineering, University of Wollongong, Wollongong, Australia, 2016.
36. Kelasidi, E.; Andrikopoulos, G.; Nikolakopoulos, G.; Manesis, S. A survey on pneumatic muscle actuators modeling. In Proceedings of the 2011 IEEE International Symposium on Industrial Electronics (ISIE), Gdansk, Poland, 27–30 June 2011; pp. 1263–1269.
37. Baughman, R.H. Carbon nanotube actuators. *Science* **1999**, *284*, 1340–1344. [CrossRef] [PubMed]
38. Mirfakhrai, T.; Jiyoung, O.; Kozlov, M.; Fok, E.C.W.; Mei, Z.; Shaoli, F.; Baughman, R.H.; Madden, J.D.W. Electrochemical actuation of carbon nanotube yarns. *Smart Mater. Struct.* **2007**, *16*. [CrossRef]
39. Lange, N.; Wippermann, F.; Leitel, R.; Bruchmann, C.; Beckert, E.; Eberhardt, R.; Tünnermann, A. First results on electrostatic polymer actuators based on uv replication. In Proceedings of the Micromachining and Microfabrication Process Technology XVI: SPIE Photonics West, San Francisco, CA, USA, 22–27 January 2011; p. 792609.
40. Johnstone, R.W.; Parameswaran, M. Electrostatic Actuators. In *An Introduction to Surface-Micromachining*; Springer: Boston, MA, USA, 2004; pp. 135–152.
41. Jones, B.E.; McKenzie, J.S. A review of optical actuators and the impact of micromachining. *Sens. Actuators A Phys.* **1993**, *37–38*, 202–207. [CrossRef]
42. Howe, D. Magnetic actuators. *Sens. Actuators A Phys.* **2000**, *81*, 268–274. [CrossRef]
43. Tiwari, R.; Meller, M.A.; Wajcs, K.B.; Moses, C.; Reveles, I.; Garcia, E. Hydraulic artificial muscles. *J. Intell. Mater. Syst. Struct.* **2012**, *23*, 301–312. [CrossRef]
44. Solano, B.; Laloy, J.; Rotinat-Libersa, C. Compact and lightweight hydraulic actuation system for high performance millimeter scale robotic applications: Modeling and experiments. In Proceedings of the ASME 2010 Conference on Smart Materials, Adaptive Structures and Intelligent Systems, Philadelphia, PA, USA, 28 September–1 October 2010; pp. 405–411.
45. Tondu, B.; Emirkhanian, R.; Mathé, S.; Ricard, A. A pH-activated using the McKibben-type braided structure. *Sens. Actuators A Phys.* **2009**, *150*, 124–130. [CrossRef]
46. Schreyer, H.B.; Gebhart, N.; Kim, K.J.; Shahinpoor, M. Electrical Activation of Artificial Muscles Containing Polyacrylonitrile Gel Fibers. *Biomacromolecules* **2000**, *1*, 642–647. [CrossRef]
47. Maziz, A.; Concas, A.; Khaldi, A.; Stålhand, J.; Persson, N.-K.; Jager, E.W.H. Knitting and weaving artificial muscles. *Sci. Adv.* **2017**, *3*. [CrossRef] [PubMed]
48. Carosio, S.; Monero, A. Smart and hybrid materials: Perspectives for their use in textile structures for better health care. *Stud. Health Technol. Inform.* **2004**, *108*, 335–343. [PubMed]

49. Abel, J.; Luntz, J.; Brei, D. A two-dimensional analytical model and experimental validation of garter stitch knitted shape memory alloy actuator architecture. *Smart Mater. Struct.* **2012**, *21*, 085011. [CrossRef]
50. Zhang, M.; Atkinson, K.R.; Baughman, R.H. Multifunctional Carbon Nanotube Yarns by Downsizing an Ancient Technology. *Science* **2004**, *306*, 1358–1361. [CrossRef]
51. Li, D.; Paxton, W.F.; Baughman, R.H.; Huang, T.J.; Stoddart, J.F.; Weiss, P.S. Molecular, supramolecular, and macromolecular motors and artificial muscles. *MRS Bull.* **2009**, *34*, 671–681. [CrossRef]
52. Mirvakili, S.M. Niobium Nanowire Yarns and their Application as Artificial Muscles. *Adv. Funct. Mater.* **2013**, *23*, 4311–4316. [CrossRef]
53. Peining, C.; Yifan, X.; Sisi, H.; Xuemei, S.; Shaowu, P.; Jue, D.; Daoyong, C.; Huisheng, P. Hierarchically arranged helical fibre actuators driven by solvents and vapours. *Nat. Nanotechnol.* **2015**, *10*, 1077–1083. [CrossRef]
54. Harun, M.H.; Saion, E.; Kassim, A.; Yahya, N.; Mahmud, E. Conjugated conducting polymers: A brief overview. *UCSI Acad. J. J. Adv. Sci. Arts* **2007**, *2*, 63–68.
55. Kaneto, K. Research Trends of Soft Actuators based on Electroactive Polymers and Conducting Polymers. *J. Phys. Conf. Ser.* **2016**, *704*, 012004. [CrossRef]
56. Foroughi, J. *Development of Novel Nanostructured Conducting Polypyrrole Fibres*; Intelligent Polymer Research Institute, Faculty of Engineering: Wollongong, Australia, 2009.
57. Madden, J. Creep and cycle life in polypyrrole actuators. *Sens. Actuators A Phys.* **2007**, *133*, 210–217. [CrossRef]
58. Kim, J. Synthesis, characterization and actuation behavior of polyaniline-coated electroactive paper actuators. *Polym. Int.* **2007**, *56*, 1530–1536. [CrossRef]
59. Xie, J. Fabrication and characterization of solid state conducting polymer actuators. *Proc. SPIE* **2004**, *5385*, 406–412. [CrossRef]
60. De Rossi, D.; Mazzoldi, A. *Linear Fully Dry Polymer Actuators*; Place of Publication: Newport Beach, CA, USA, 1999; pp. 35–44.
61. Takashima, W. The electrochemical actuator using electrochemically-deposited poly-aniline film. *Synth. Met.* **1995**, *71*, 2265–2266. [CrossRef]
62. Simaite, A. Towards inkjet printable conducting polymer artificial muscles. *Sens. Actuators B Chem.* **2016**, *229*, 425–433. [CrossRef]
63. Okuzaki, H. Ionic liquid/polyurethane/PEDOT:PSS composites for electro-active polymer actuators. *Sens. Actuators B Chem.* **2014**, *194*, 59–63. [CrossRef]
64. Wang, G. Actuator and Generator Based on Moisture-Responsive PEDOT: PSS/PVDF composite film. *Sens. Actuators B Chem.* **2018**, *255*, 1415–1421. [CrossRef]
65. Naficy, S. Evaluation of encapsulating coatings on the performance of polypyrrole actuators. *Smart Mater. Struct.* **2013**, *22*, 075005. [CrossRef]
66. Fengel, C.V.; Bradshaw, N.P.; Severt, S.Y.; Murphy, A.R.; Leger, J.M. Biocompatible silk-conducting polymer composite trilayer actuators. *Smart Mater. Struct.* **2017**, *26*, 055004. [CrossRef]
67. Khaldi, A.; Maziz, A.; Alici, G.; Spinks, G.M.; Jager, E.W.H. Bottom-up microfabrication process for individually controlled conjugated polymer actuators. *Sens. Actuators B Chem.* **2016**, *230*, 818–824. [CrossRef]
68. Burriss, E.T.; Alici, G.; Spinks, G.M.; McGovern, S. Modelling and Performance Enhancement of a Linear Actuation Mechanism Using Conducting Polymers. *Inf. Control Autom. Rob.* **2011**, *85*, 63–78.
69. Spinks, G.; Binbin, X.; Campbell, T.; Whitten, P.; Mottaghitalab, V.; Samani, M.B.; Wallace, G.G. In pursuit of high-force/high-stroke conducting polymer actuators. In Proceedings of the Volume 5759, Smart Structures and Materials 2005: Electroactive Polymer Actuators and Devices (EAPAD), San Diego, CA, USA, 6 May 2005; pp. 314–321.
70. Baughman, R.H. Conducting polymer artificial muscles. *Synth. Met.* **1996**, *78*, 339–353. [CrossRef]
71. Madden, J.D.; Madden, P.G.; Anquetil, P.A.; Hunter, I.W. Load and time dependence of displacement in a conducting polymer actuator. *Mat. Res. Soc. Symp. Proc.* **2002**, *698*, 137–144.
72. Anquetil, P.A.; Rinderknecht, D.; Vandesteeg, N.A.; Madden, J.D.; Hunter, I.W. Large strain actuation in polypyrrole actuators. In Proceedings of the Smart Structures and Materials 2004: Electroactive Polymer Actuators and Devices (EAPAD), San Diego, CA, USA, 27 July 2004; pp. 380–387.
73. Madden, J.D. Encapsulated polypyrrole actuators. *Synth. Met.* **1999**, *105*, 61–64. [CrossRef]

74. Farajollahi, M.; Woehling, V.; Plesse, C.; Nguyen, G.T.M.; Vidal, F.; Sassani, F.; Yang, V.X.D.; Madden, J.D.W. Self-contained tubular bending actuator driven by conducting polymers. *Sens. Actuators A Phys.* **2016**, *249*, 45–56. [CrossRef]
75. Lee, J.A.; Li, N.; Haines, C.S.; Kim, K.J.; Lepró, X.; Ovalle-Robles, R.; Kim, S.J.; Baughman, R.H. Electrochemically Powered, Energy-Conserving Carbon Nanotube Artificial Muscles. *Adv. Mater.* **2017**, *29*, 1700870. [CrossRef]
76. Lee, J.A.; Kim, Y.T.; Spinks, G.M.; Suh, D.; Lepró, X.; Lima, M.D.; Baughman, R.H.; Kim, S.J. All-Solid-State Carbon Nanotube Torsional and Tensile Artificial Muscles. *Nano Lett.* **2014**, *14*, 2664–2669. [CrossRef]
77. Chu, H.-Y. Microsystems. "CNT-Polymer" Composite-Film as a Material for Microactuators. In Proceedings of the IEEE Transducers 2007 International Solid-State Sensors, Actuators and Microsystems Conference, Lyon, France, 10–14 June 2007; pp. 1549–1552.
78. Dang, D.X.; Truong, T.K.; Lim, S.C.; Suh, D. Multi-dimensional actuation measurement method for tensile actuation of paraffin-infiltrated multi-wall carbon nanotube yarns. *Rev. Sci. Instrum.* **2017**, *88*, 075001. [CrossRef]
79. Chen, L.Z.; Liu, C.H.; Hu, C.H.; Fan, S.S. Electrothermal actuation based on carbon nanotube network in silicone elastomer. *Appl. Phys. Lett.* **2008**, *92*, 263104. [CrossRef]
80. Srivastava, S.; Bhalla, S.; Madan, A. A review of rotary actuators based on shape memory alloys. *J. Intell. Mater. Syst. Struct.* **2017**, *28*, 1863–1885. [CrossRef]
81. Lan, C.-C.; Wang, J.-H.; Fan, C.-H. Optimal design of rotary manipulators using shape memory alloy wire actuated flexures. *Sens. Actuators A Phys.* **2009**, *153*, 258–266. [CrossRef]
82. Andrianesis, K.; Koveos, Y.; Nikolakopoulos, G.; Tzes, A. Experimental study of a shape memory alloy actuation system for a novel prosthetic hand. In *Shape Memory Alloys*; InTech: London, UK, 2010.
83. Mohd Jani, J.; Leary, M.; Subic, A. Designing shape memory alloy linear actuators: A review. *J. Intell. Mater. Syst. Struct.* **2017**, *28*, 1699–1718. [CrossRef]
84. Luo, H.; Liao, Y.; Abel, E.; Wang, Z.; Liu, X. Hysteresis behaviour and modeling of SMA actuators. In *Shape Memory Alloys*; InTech: London, UK, 2010.
85. Stegmaier, T.; Mavely, J.; Schneider, P. CHAPTER 6: High-Performance and High-Functional Fibres and Textiles. In *Textiles in Sports*; Elsevier: Amsterdam, The Netherlands; pp. 89–119.
86. Choy, C.L.; Chen, F.C.; Young, K. Negative thermal expansion in oriented crystalline polymers. *J. Polym. Sci. Polym. Phys. Ed.* **1981**, *19*, 335–352. [CrossRef]
87. Steele, J.R.; Gho, S.A.; Campbell, T.E.; Richards, C.J.; Beirne, S.; Spinks, G.M.; Wallace, G.G. The Bionic Bra: Using electromaterials to sense and modify breast support to enhance active living. *J. Rehabil. Assist. Technol. Eng.* **2018**, *5*. [CrossRef]
88. Sárosi, J.; Csikós, S.; Asztalos, I.; Gyeviki, J.; Véha, A. Accurate Positioning of Spring Returned Pneumatic Using Sliding-mode Control. In Proceedings of the 1st Regional Conference—Mechatronics in Practice and Education MECH (CONF 2011), Subotica, Serbia, 8–10 December 2011.
89. Klute, G.K.; Czerniecki, J.M.; Hannaford, B. Artificial muscles: Actuators for biorobotic systems. *Int. J. Robot. Res.* **2002**, *21*, 295–309. [CrossRef]
90. Janickis, V.; Ancutienė, I. Modification of polyester textile by conductive copper sulfide layers. *Poliesterinio Audinio Modifikavimas Elektrai Laidžiais Vario Sulfidų Sluoksniais* **2009**, *20*, 136–140.
91. Bashir, T.; Ali, M.; Persson, N.-K.; Ramamoorthy, S.K.; Skrifvars, M. Stretch sensing properties of conductive knitted structures of PEDOT-coated viscose and polyester yarns. *Text. Res. J.* **2014**, *84*, 323–334. [CrossRef]
92. Maity, S.; Chatterjee, A.; Singh, B.; Pal Singh, A. Polypyrrole based electro-conductive textiles for heat generation. *J. Text. Inst.* **2014**, *105*, 887–893. [CrossRef]
93. Lin, T.; Wang, L.; Wang, X.; Kaynak, A. Polymerising pyrrole on polyester textiles and controlling the conductivity through coating thickness. *Thin Solid Films* **2005**, *479*, 77–82. [CrossRef]
94. Zhang, B.; Xue, T.; Meng, J.; Li, H. Study on property of PANI/PET composite conductive fabric. *J. Text. Inst.* **2014**, *106*, 253–259. [CrossRef]
95. Molina, J.; Zille, A.; Fernández, J.; Souto, A.P.; Bonastre, J.; Cases, F. Conducting fabrics of polyester coated with polypyrrole and doped with graphene oxide. *Synth. Met.* **2015**, *204*, 110–121. [CrossRef]
96. Bhattacharya, R.; Pieterson, L.V.; Os, K.V. Improving conduction and mechanical reliability of woven metal interconnects. *IEEE Trans. Compon. Packag. Manuf. Technol.* **2012**, *2*, 165–168. [CrossRef]

97. Bashir, T.; Skrifvars, M.; Persson, N.K. Surface modification of conductive PEDOT coated textile yarns with silicone resin. *Mater. Technol.* **2011**, *26*, 135–139. [CrossRef]
98. Xuyuan, T.; Koncar, V.; Tzu-Hao, H.; Chien-Lung, S.; Ya-Chi, K.; Gwo-Tsuen, J. How to Make Reliable, Washable, and Wearable Textronic Devices. *Sensors* **2017**, *17*, 1–16. [CrossRef]
99. Mondal, S. Phase change materials for smart textiles—An overview. *Appl. Therm. Eng.* **2008**, *28*, 1536–1550. [CrossRef]
100. Saito, Y.; Suzuki, Y.; Daitoku, K.; Minakawa, M.; Fukuda, I.; Goto, T. Cardiac supporting device using artificial rubber muscle: Preliminary study to active dynamic cardiomyoplasty. *J. Artif. Organs* **2015**, *18*, 377–381. [CrossRef]
101. Sherif, H.M.F. The artificial ventricle: A conceptual design for a novel mechanical circulatory support system. *Minim. Invasive Ther. Allied Technol.* **2009**, *18*, 178–180. [CrossRef]
102. Ruhparwar, A.; Piontek, P.; Ungerer, M.; Ghodsizad, A.; Partovi, S.; Foroughi, J.; Szabo, G.; Farag, M.; Karck, M.; Spinks, G.M.; et al. Electrically Contractile Polymers Augment Right Ventricular Output in the Heart. *Artif. Organs* **2014**, *38*, 1034–1039. [CrossRef]

© 2019 by the authors. Licensee MDPI, Basel, Switzerland. This article is an open access article distributed under the terms and conditions of the Creative Commons Attribution (CC BY) license (http://creativecommons.org/licenses/by/4.0/).

Review

Review on Fabrication of Structurally Colored Fibers by Electrospinning

Jiali Yu and Chi-Wai Kan *

Institute of Textiles and Clothing, The Hong Kong Polytechnic University, Hung Hom, Kowloon, Hong Kong, China; 17901377r@connect.polyu.hk
* Correspondence: tccwk@polyu.edu.hk; Tel.: +852-2766-6531

Received: 29 June 2018; Accepted: 20 September 2018; Published: 26 September 2018

Abstract: Structural color derived from the physical interactions of photons, with the specific chromatic mechanism differing from that of dyes and pigments, has brought considerable attention by the conducive virtue of being dye-free and fadeless. This has recently become a research hot-spot. Assemblies of colloidal nanoparticles enable the manufacture of periodic photonic nanostructures. In our review, the mechanism of nanoparticle assemblies into structurally colored structures by the electrospinning method was briefly introduced, followed by a comparatively comprehensive review summarizing the research related to photonic crystals with periodically aligned nanostructures constructed by the assembly of colloidal nanoparticles, and the concrete studies concerning the fabrication of well-aligned electrospun nanofibers incorporating with colloidal nanoparticles based on the investigation of relevant factors such as the sizes of colloidal nanoparticles, the weight ratio between colloidal nanoparticles, and the polymer matrix. Electrospinning is expected to be a deserving technique for the fabrication of structurally colored nanofibers while the colloidal nanoparticles can be well confined into aligned arrangement inside nanofibres during the electrospinning process after the achievement of resolving remaining challenges.

Keywords: structural color; electrospun nanofibers; nanoparticle assemblies; polymer; photonic crystal

1. Introduction

Structural color arises from photonic crystals possessing a periodic modulation assembled by colloidal nanoparticles; this has invited great interest since their introduction in the late 1980s [1–3]. The different mechanisms for controlling and manipulating light provided by the self-assembly of photonic crystals could be analyzed on the basis of the photonic bandgap [4]. Light refraction, diffraction, and scattering exist in the periodic nanostructures that contain regularly repeating internal regions [5]. Colloidal nanoparticles used to assemble photonic crystals with closely-packed structures play an important role in generating structural color to act on the transport and manipulation of light, owing to the Bragg diffraction [6]. Photonic crystals that exhibit brilliant structural colors are usually prepared by self-assembly of inorganic colloidal nanoparticles like silicon dioxide or silica (SiO_2) [7,8], titanium dioxide (TiO_2), core-shell colloids containing metallic nanoparticles like gold (Au)@SiO_2 colloids [9], polymeric nanoparticles including colloidal homopolymer nanoparticles [10,11], polystyrene (PS) [8,12,13], and copolymer nanoparticles like poly(styrene-methyl methacrylate-acrylic acid) [14]. Electrospinning is one of the favorable fabrication techniques that utilizes electrostatic forces to prepare polymer filament networks as scaffolds in nanometer scale, since it emerged in the early 1930s [15]. The periodically arranged nanostructures constructed by electrospinning have shown the potential of generating brilliant colors with the addition of photonic crystals, leading to a wide range of prospects in a novel biomimetic field.

In terms of the electrospinning technique, with the advantages of easy access to the adjustment of morphology of nanofibres and fast continuous preparation process [16], various kinds of electrospun nanofiber assemblies with desired properties could be prepared for myriad specific applications [17]. Unlike the dip-coating method [18], the electrospun nanofibers could be prepared without adhesion between the substrate and coating layer. No high thermal fulfillment tends to be required for the fabrication process, which is dissimilar to the melting and shear ordering method [19–21]. Electrospinning possesses high efficiency of manufacturing than the combined approach of microfluidic emulsification and solvent diffusion [22]. The electrophoretic method [23–25] has served as a fabrication technique for structurally colored fiber that has been examined on colloidal crystal growth with ordered arrangement for several decades, but the limitation of the substrate requirement with good electrical conductivity to form functional nanostructures cannot be neglected. Accordingly, electrospinning is conducted as a more effective approach for the production of structurally colored nanofibers of polymers compared with such template-directed assembly methods in which the subsequent removal of the sacrificial templates requires a troublesome and time-consuming process [26].

Even through structural color has been introduced into colloidal systems for a long time, researches regarding structurally colored electrospun nanofibers in combination with the assembly of nanoparticles are still quite sparse. Herein, this paper reviews highly relevant and significant research on structurally colored fibers with periodically aligned nanostructures by a promising electrospinning technique.

2. Mechanism of Nanoparticle Assemblies into Structurally Colored Nanofibers

While propagating the periodic structures, the wavelength of reflected light on the nanoparticle-assembled nanofibers is predicted by a Bragg equation [27].

$$\lambda = 2dn_{eff} = (8/3)^{1/2} D_p (0.74 n_p^2 + 0.26 n_m^2 - \sin^2\theta)^{1/2} \tag{1}$$

where, λ, d, n_{eff}, D_p, n_m, n_p, θ refer to light wavelength of the reflected color, the periodicity of lattice spacing, effective refractive index, nanoparticle diameter, the refractive index of polymer matrix, the refractive index of nanoparticle, and light incidence angle, respectively.

If the light wavelength that fulfills the Bragg condition, incident light would undergo diffraction and hence the color of specific reflected light would appear on the surface of the ordered structures as demonstrated in Figure 1. The periodically ordered structures with aligned nanofibers are established by electrospinning, in which the round spheres represent the nanoparticles while the polymer matrix that surrounds the nanoparticles is mainly removed. Inserted image means cross section of part of the aligned nanofibers within a range around several microns made of nanoparticles. The periodicity and sizes of nanoparticles closely correlate to the color of reflected light, which is mainly controlled by modulating electrospinning parameters and emulsion polymerization conditions, respectively. Structural colors primarily depend on morphology of the material with certain properties, such as refraction index, playing a significant role of determining the color as well as color intensity [28]. The differences between density in amorphous and crystalline phases of the same polymer lead to differences in the refractive index, which means the ratio of the velocity of light in a vacuum to that in the polymer is closely related to the generation of structural color due to the significant role of polymers in the development of materials for photonics [6]. The width and frequency of the reflection tend to be determined by refractive index as well as the layer or film thickness in the photonic crystal film [7]. Hence refractive index has been identified as one of the fundamental optical properties of a polymer that is regarded as an input parameter for assessing many other optical properties [29,30]. A readily feasible method for changing color is to modulate the refractive indices of the nanostructures constructed by polymers [31]. Enlarging the difference between refractive index and colloidal nanoparticles and polymer binder tends to be accessible to strength the intensity of structural color. The surface polarity between the photonic crystal nanoparticles and polymer binder is shown to exert great significance in the preparation of spinning solutions for establishing electrospun nanofibers, and subsequently it would be routinely considered while producing the electrospinning solution.

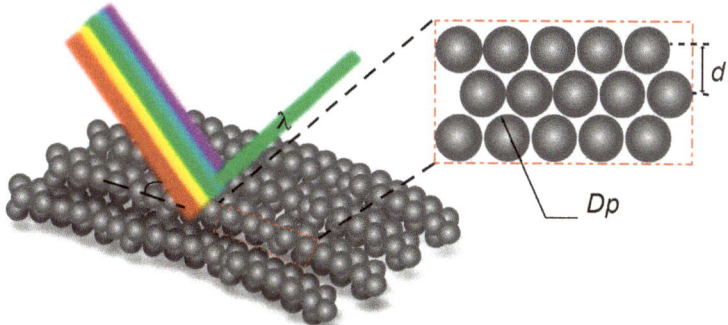

Figure 1. Demonstration of generating structural color according to Bragg's law.

A schematic representation of the preparation of nanoparticle assemblies into nanofibres with photonic crystal structures exhibiting brilliant color is given in Figure 2, which is similar to the given work [32]. Like conventional electrospinning setup, a spinneret with needle and syringe, a high voltage power supply, and a conductive collector including rotating drum (Figure 2a) or parallel collector (Figure 2b) are constructed for manufacturing well-aligned nanofibres, and the working principle of electrospinning is similar to the previous studies [33]. Syringe pump filled with spinning solution is used to provide the liquid to the tip of the needle in a certain flow rate. Due to the surface tension, a liquid droplet could form, where charges would generate while imposing a high voltage at the spinneret. A Taylor cone would exist when charge repulsion is strong enough to conquer the surface tension. Then the liquid could be ejected to the collector under the driving forces based on the electrostatic repulsion. Finally, the nanofibers align perpendicular to the axis of the drum or parallel electrode with the evaporation of solvent [34]. The mechanism that modeled and simulated the corresponding electrostatic field distributions between the needle and the collector in a three-dimensional (3D) space is shown in Figure 2c,d. The typical target collectors utilize a rotating drum and parallel collector, respectively. The electrostatic driving forces have no fixed direction that can be ascribed to the unstable electric field. In order to obtain the highly aligned nanofibres fabricated by electrospinning [34,35], it has been found to apply high rotating collectors, parallel electrode collectors [36,37], and auxiliary electrodes [38,39]. The gap width between the as-spun nanofibers reached the values ranging from centimeters to several micrometers as investigated in [40], which resulted in the uncontrollable orientation of electrospun nanofibers, and hence the deep investigation of suitable electrospinning conditions is required to build the scaffolding of aligned nanostructures that exhibits structural color. For instance, Li et al. [40] have analyzed the electrostatic forces using continuous conductive collector without and with an insulating gap, they found that a parallel collector had the priority of enabling the nanofibers in an uniaxial alignment.

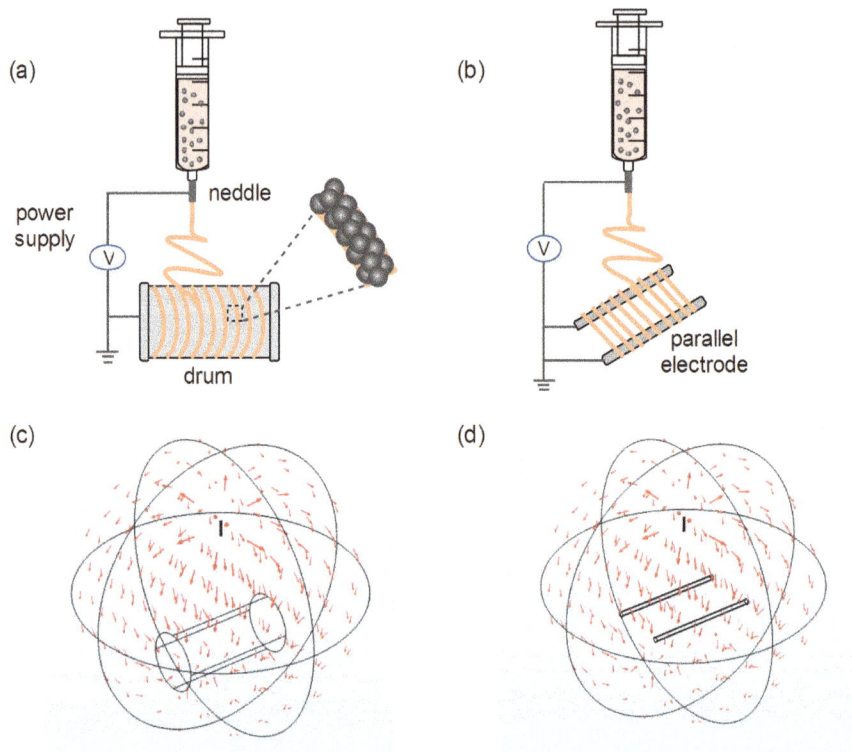

Figure 2. Schematic diagram of electrospinning setup using (**a**) rotating drum, (**b**) parallel electrode; representative simulation of electrostatic field distribution between the needle and (**c**) rotating drum, (**d**) parallel electrode using COMSOL software. Red arrows refer to electrostatic forces.

3. Photonic Crystals for Fabricating Structurally Colored Fibers

The construction of photonic crystals with three-dimensional periodic geometry is a crucial part when it comes to fabricating structurally colored fibers. Coherently diffraction and scattering by the photons of specific wavelengths could occur on the periodic lattice planes of photonic crystals, which has a spacing that is similar to such wavelengths [28,41]. Prior to the preparation of three-dimensional photonic crystal structural color materials, the self-assembly of colloidal nanoparticles of spherical or cylindrical shapes into a close-packed and periodic array tends to enable the photonic crystal as an attractive candidate for generating structurally colored fibers. Colloidal nanoparticles can act as building blocks for constructing photonic crystals as a result of the superiority that includes multiple choices of chemical composition, cost-effective products, as well as tunable particle sizes [5]. The size dimension of nanoparticles mainly controlled by the adjustment of the reaction conditions during synthesis process, including the different amounts of reactants [42,43], precursor type, reaction temperature, and time [44], causes a shift of color that enlarges the potential applications of promising areas.

3.1. Inorganic Nanoparticles for Photonic Crystals

Since Iler [45] first carried out the research that multilayers assembled by colloidal nanoparticles could be formed on rigid substrates by the layer-by-layer (LBL) deposition method, many researchers have studied different nanoparticles, including inorganic nanoparticles for constructing photonic crystals, such as silicon dioxide or silica (SiO$_2$) and titanium dioxide (TiO$_2$), which are the two preferred materials for constructing a desired array with uniform arrangement because of the different dielectric constants they have. The controllable synthesis process of monodisperse silica spheres with micron sizes was presented by Stöber in the1960s, which has been used as a widespread reaction [46]. Wang et al. [7] adopted a new strategy to fabricate color-tunable biomimetic film with tunable structural colors in organic solvent by assembling SiO$_2$ nanospheres, and the experimental results indicated that carbon black dopant remarkably increases the chroma and enhances the intensity of the structural colors. Structural color changes of colloidal crystal films from red to violet were determined by the assembly of SiO$_2$ nanoparticles with different particle sizes from 207 to 350 nm using the Stöber process [43]. A recent study by Zhang [47] disclosed vivid structural color coatings from atomization deposition of SiO$_2$ nanoparticles with poly (vinyl alcohol) additive on silk fabrics, which can possess robust mechanical properties as depicted in Figure 3.

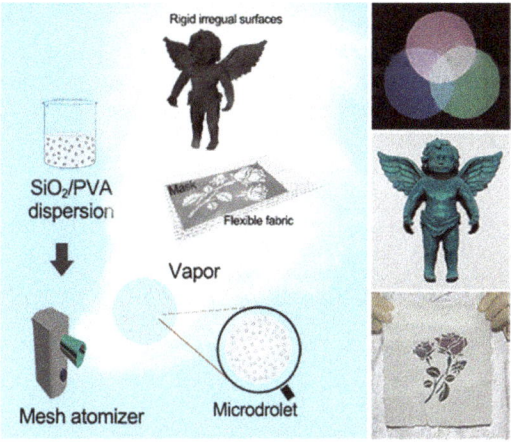

Figure 3. Schematic illustration showing the fabrication process of amorphous SiO$_2$ nanoparticles into photonic crystals with noniridescent colors on various substrate materials by atomization deposition. Reprinted with permission from [47] American Chemical Society, 2018.

Several conventional techniques have been employed to produce TiO$_2$ films, including conventional sol-gel method [48,49], vapor deposition method [50,51], and liquid phase deposition [52,53]. Chen and co-workers [54] successfully used an effective method named the atomic layer deposition (ALD) technique to fabricate TiO$_2$ coatings with vibrant and uniform structural colors using black carbon fibers as substrates, showing excellent laundering durability. These have many potential applications in the field of optical and color-display devices. Because the use of single inorganic particles is significantly limited to the formation of periodically arranged photonic structures, core-shell colloidal particles have become desirable materials. Metal nanoparticles like gold (Au) or silver (Ag) are extraordinarily efficient at strongly absorbing and scattering light to enhance the intension of reflected light for exhibiting brilliant color, due to the optical property known as localized surface plasmon resonances [55,56]. SiO$_2$ nanoparticles incorporated with gold has been prepared to form colloidal spheres (Au@SiO$_2$) with core-shell shapes, which can serve as the building blocks for photonic applications [9]. Yuan et al. [50] reported structural colors on polyester fabric with the coating of silver(Ag)/TiO$_2$ films made by magnetron sputtering. The resultant samples with diverse colors of purple, light blue, blue, pink, and dark red could be displayed by controlling different thickness of TiO$_2$ thin films. Luo et al. [57] reported electrically induced structural color alteration of core-shell (SiO$_2$@TiO$_2$) photonic crystals for electric field applications.

3.2. Polymeric Nanoparticles for Photonic Crystals

Since the discovery of an approach for synthesizing monodisperse polymer colloids by Vanderhoff, this synthetic method has been used to prepare diverse polymeric colloids, including homopolymer colloidal nanoparticles as well as copolymer nanoparticles, besides inorganic colloidal nanoparticles [58,59]. These monodisperse polymeric colloidal nanoparticles prepared by some mature synthesis techniques, such as emulsion polymerization, dispersion polymerization, precipitation polymerization, or seeded polymerization, can be incorporated into polymer solution for electrospinning [5]. Among the synthetic routes, emulsion and dispersion polymerization are the most commonly used.

Colloidal homopolymer nanoparticles such as polymethylmethacrylate and polystyrene monodispersed spheres, have been successfully synthesized to be used for fabricating structural color. The size of such nanoparticles influencing optical properties of electrospun nanostructures can be determined by synthesis conditions like concentration of initiator, reaction time, and temperature. The preparation of nanofibres containing polymeric nanoparticles by electrospinning has been examined [12]. Meng et al. [60] studied the mechanical stability and hydrophobicity of structural color films composed of PS microspheres, which was of great significance for potential architectural purposes in paint and decoration. According to Han et al. [61], the stability of switching performance of the tunable photonic crystals consisting of well-ordered PS colloidal arrays has been improved in long range for display applications. Tang and co-workers [62,63] presented a kind of polymer opal film with brilliant structural colors from PS nanospheres or crosslinked PS/polydimethylsiloxane (PDMS) with well-aligned structures by using the self-assembly method with thermal assistance for architectural applications. Gu et al. [8] described a kind of structural color with a lotus effect formed by mixing both PS spheres and SiO_2 particles to deposit an opal film on a glass substrate. Meanwhile, researchers have made much efforts to prepare poly(methyl methacrylate) PMMA nanospheres for many promising purposes, synthesized by a commonly used method called emulsion polymerization [10,64,65]. Tang et al. [66] developed a kind of heat-resistant photonic crystal film consisting of PMMA colloidal spheres with different colors by investigating the effect of methyl methacrylate (MMA) concentration on particle size as well as the distribution of PMMA particles.

Despite colloidal homopolymer nanoparticles, a variety of copolymer nanospheres commonly exhibiting core-shell nanostructures can be synthesized with other kinds of monomers by polymerization to achieve the desired functions. Liu et al. [67] developed a kind of photonic crystal named poly(styrene-methacrylic acid) coating on the textile fabrics exhibiting brilliant and variable structural colors, possessing better hydrophobic properties. Non-iridescent brilliant structural colors (blue, green and red) from photonic structures based on assembling polystyrene/poly(N-isopropylacrylamide-co-acrylic-acid) particles with core-sheath particle structure over the full-spectrum were demonstrated by Park [68], as displayed in Figure 4a,b.

Figure 4c,d shows that a kind of core-sheath colloidal poly(styrene-methyl methacrylate-acrylic acid) nanospheres was synthesized by emulsion polymerization using three monomers including methyl methacrylate, styrene and acrylic acid from [14]. Brilliant and monochromatic colors of colloidal crystal films covering the entire visible range from red to violet color on a glass substrate were obtained by assembling such nanospheres with various diameters. Yuan et al. [69] also successfully synthesized monodispersed poly (styrene-methyl methacrylate-acrylic acid) composite nanospheres and it was designed to be blended with a measured amount of PVA to produce a dye-free electrospun fibrous membrane with tunable brilliant structural colors by colloidal electrospinning. The schematic illustration of this process and the typical scanning electron microscope (SEM) images of the nanofibres are depicted in Figure 5. The SEM results demonstrated that the colloidal spheres are well-arranged on the surfaces of nanofibres in a cylindrical shape.

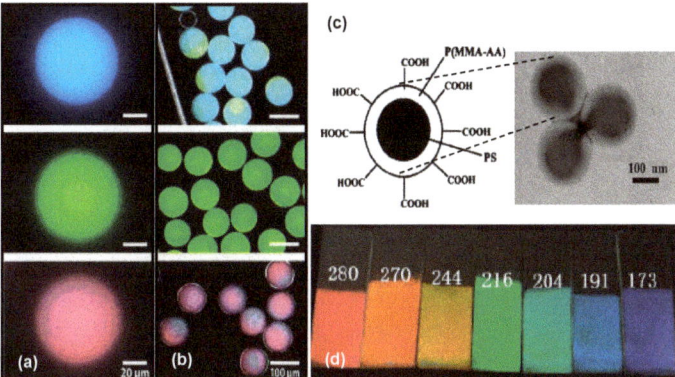

Figure 4. (**a**,**b**) Three structural colors (blue, green, and red) of photonic microcapsules prepared with various shell thicknesses of core-sheath particles under optical micrographs. (**a**) bright-field, (**b**) dark-field. Reprinted with permission from [68] Wiley-VCH Verlag GmbH & Co. KGaA, Weinheim, 2014; (**c**) Schematic diagram of poly(St-MMA-AA) nanostructure and the typical transmission electron microscope (TEM) image of the core-sheath nanospheres with the diameter of 173 nm; (**d**) Pictures of prepared films with different sizes deposited on the glass substrate. Reprinted with permission from [14] Wiley-VCH Verlag GmbH & Co. KGaA, Weinheim, 2006.

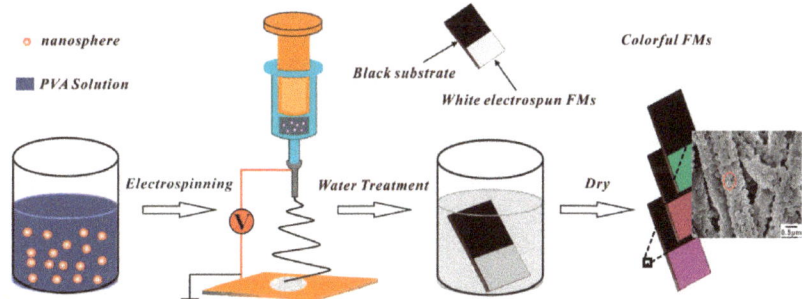

Figure 5. Schematic illustration of the fabrication process of electrospinning for preparation of colorful fibrous membranes. Reprinted with permission from [69] American Chemical Society, 2015.

4. Electrospun Nanofibers Derived from Photonic Colloidal Nanoparticles

Many researchers have successfully achieved structural color films or photonic opals or coatings on substrates [43,70–72], and the electrospinning technique for generating common nanofibres has been proven. Electrospinning using a high-voltage electric field to spin polymer solutions into nanofibres with nanoscale diameters, tends to be a versatile technique for successfully manufacturing polymer nanofibres from various kinds of polymer materials, like poly (vinyl alcohol), poly(ethylene terephthalate), poly(vinyl phenol), polyurethanes, polyamide, poly(methyl methacrylate), polyacrylonitrile, poly (ether imide), poly (ethylene gricol) [73], which are treated as a polymer matrix, an indispensable element and the main component of the nanostructures exhibiting brilliant colors. More importantly, the fabrication of nanofiber membranes containing SiO_2 or TiO_2 colloidal nanoparticles by the electrospinning method has proven to be feasible according to previous works by Zhang [74], Im [75] and Pant [76]. A discovery of present work focusing on incorporating colloidal nanoparticles into electrospun nanofibers with well aligned morphology has been found as shown in Table 1 with detailed parameters. Different morphologies of electrospun nanofibers could be achieved from random packed, necklace-like to cylindrically hexagonal patterned structures by controlling several decisive conditions, which covers the different properties of the polymer matrix, including crystallinity, glass transition temperature, molecular weight, and spinning solution properties including surface tension, viscosity, conductivity, polymer solubility as well as the electrospinning processing parameters consisting of the feeding rate, applied voltage, receiving distance, effect of collector, and needle [15]. Besides, with the addition of photonic colloidal nanoparticles, more influencing factors are thought to be considered. The sizes and shapes of colloidal nanoparticles blended in polymer matrix is one of the indispensable factors determining the morphology of the as-prepared electrospun nanofibers. In the early 20th century, Yang et al. [26] presented the concept that colloidal particles can be confined and assembled inside nanofibres during the electrospinning process and they successfully fabricated polyacrylamide/SiO_2, poly(ethyleneoxide)/SiO_2 and poly(acrylonitrile)/SiO_2 nanofibers with close-packed SiO_2 particles formed in radial direction of the nanofibres. From SEM results in Figure 6, it can be documented that with the increasing size of SiO_2 particles from 100 nm to 1000 nm, the structure of polyacrylamide nanofibres transformed from less-aligned to assembled necklace-like structures.

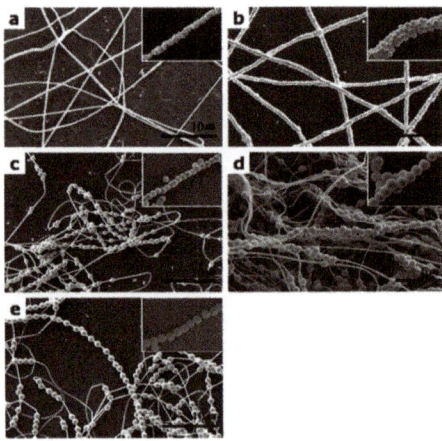

Figure 6. Scanning electron microscope (SEM) images of the structures and morphology of polyacrylamide nanofibres with the increase of SiO_2 nanoparticle sizes: (**a**) 100 nm, (**b**) 300 nm, (**c**) 450 nm, (**d**) 700 nm, and (**e**) 1000 nm. Reprinted with permission from [26] American Chemical Society, 2006.

Another affecting factor is the weight ratio between photonic colloidal nanoparticles and polymer matrix. Yuan and his group members [12] investigated the morphology of polystyrene nanosphere(PS)/poly(vinyl alcohol)(PVA) electrospun fibers with the various structures. Polystyrene nanospheres were synthesized by emulsion polymerization using the monomer (styrene) with the addition of an emulsifier. The results showed that string-on-bead and necklace-like nanofibres were obtained with certain ratios of PS and PVA, as shown in Figure 7e. A common electrospinning factor named polymer concentration was also analyzed as can be seen in Figure 7g–n.

Despite the construction for morphology structures of electrospun nanofibers, extra factors may affect the color on the nanofiber surfaces according to Table 1. To evaluate the reason that most of the resulting nanofibers with uniform ordered structures were achromatous, the researchers found that it was attributed to the low refractive index. The subsequent exploration verified that after establishing the scaffolding in which electrospun nanofibers were in alignment, the polymer matrix could be removed without the destruction of the desired nanofiber structures that was conducive to yielding color. The dissolution of the PVA matrix of poly (styrene-methyl methacrylate-acrylic acid)/PVA nanofibers increased the refractive index contrast resulted in displaying structural color [69].

Table 1. Fundamental parameters of nanoparticles contained electrospun nanofibers with well-aligned morphology.

Colloidal Nanoparticles	Sizes (nm)	C_1 (wt%)	Polymer Matrix	Mw	C_2 (wt%)	R	Rf (mL/h)	V (kv)	D (cm)	T; H	Morphology	Ref.
silica	15, 50, 100	20	polyvinyl alcohol	Mn: 66,000	10	2:3	1	10	10	25 °C 50 ± 5%	grain-like singly aligned	[77]
silica	100, 300, 450, 700	20	Polyacrylamide poly(ethyleneoxide)	600,000–1,000,000 600,000	10	2:3	0.5–3	5–13	10	500 °C calcination	necklace-like	[26]
silica	143, 265, 910	12.2 21.6 13.1	polyvinyl alcohol	88,000	12 10 12	500:500 600:400 200:800 300:700	—	10–30	10	—	necklace-like blackberry-like	[32]
Silica polystyrene	700, 50 237	34 10	Polyacrylamide poly(ethyleneoxide)	600,000–1,000,000 600,000	10	—	0.5	5–13	10	calcination	stand-alone structures, superhydrophobic	[78]
polystyrene	100, 200, 335	40	polyvinyl alcohol	145,000 195,000	6	80:20	0.7	0–30	20	15-18 °C 30–50%	random to relative compact packing	[79]
polystyrene	225, 473	10, 15, 20, 30, 40	polyvinyl alcohol	14,500	13	1:1 2:1 4:1	0.2–0.8	10	15	25 °C 50 ± 5%	blackberry-like to uniform	[12]
poly (styrene-methyl methacrylate-acrylic acid)	220, 246, 280	40	polyvinyl alcohol	14,500	13	4:1	0.5	10	15	—	green, red, purplish-red color cylindrically hexagonal ordered	[69]
poly (N-isopropylacrylamide-co-tert-butyl acrylate)	226	40	polyacrylamide	146,000	16	1:4 2:3 1:1 3:2 4:1	0.5	10	15	20 °C 30 ± 5%	necklace-like blackberry-like	[16]

Where, Mw and Mn are the weight-average and number-average molecule weight of polymer, respectively; C_1, C_2 denote the concentration of nanoparticles and polymer matrix, respectively; R is the ratio of nanoparticle and polymer; Rf is the feed rate; V is electrospinning applied voltage; D means the distance between needle and collector; T, H represent temperature and humidity. Rf, V, D, T, H represent the parameters in the electrospinning process.

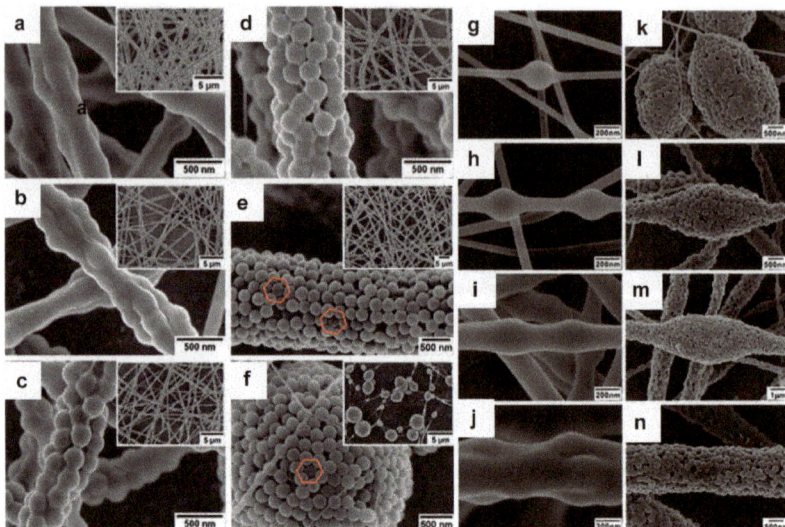

Figure 7. SEM graphs of polystyrene nanoparticles into PVA nanofibres: (**a–f**) different weight ratios of PS and PVA of 1:4, 1:2, 1:1, 2:1, 4:1 and 6:1; The inserted images in each subfigure from a to f are measured under lower magnification. (**g–n**) different concentrations of PVA: 7, 9, 11, 13 wt%, while (**g–j**) PS:PVA is 1:4 and (**k–n**) PS:PVA is 4:1. Reprinted with permission from [12] American Chemical Society, 2012.

5. Outlook and Conclusions

Myriad studies have successfully produced well-ordered nanostructures on electrospun nanofibers with the blending of photonic nanoparticles, however, the research on nanofibres containing photonic nanoparticles endowing homogeneous and non-iridescent structural colors remains very few, which indicates that it seems to be difficult to achieve the well-ordered structure of electrospun nanofibers incorporating photonic nanoparticles as a result of the overall consideration of a multitude of factors and thereby it remains a daunting task.

Overall, tremendous achievements in structural coloration based on photonic crystals have been visible in many application areas, especially optical fields, but so far only a relatively small number of researchers have tried to prepare nanofibres with brilliant structural colors by electrospinning because of the difficult regulation of manufacturing conditions. Major efforts are still required to further develop and improve effective manufacturing approaches to structurally colored nanofibers. Researchers have presented electrospun nanofibers exhibiting vivid structural colors with the addition of polymeric particle colloids, however, inorganic spheres with higher reflective indices are expected be used to fabricate structurally colored fibers because the high refractive index contrast is beneficial for increasing the intensity of structural colors. Other properties of structurally colored fibers such as mechanical properties are supposed to be considered and examined for more promising applications. It could be concluded that electrospinning is a versatile approach that could be used for the generation of highly aligned electrospun nanofibers, showing the promise of paving the way for a revolution in the development of structurally colored materials for various dye-free applications.

Author Contributions: Conceptualization, J.Y. and C.-W.K.; Data curation, J.Y.; Formal analysis, J.Y. and C.-W.K.; Funding acquisition, C.-W.K.; Investigation, J.Y. and C.-W.K.; Methodology, J.Y. and C.-W.K.; Project administration, C.-W.K.; Resources, C.-W.K.; Supervision, C.-W.K.; Validation, J.Y. and C.-W.K.; Visualization, J.Y.; Writing—original draft, J.Y.; Writing—review & editing, C.-W.K.

Funding: This work was funded by The Hong Kong Polytechnic University with grant numbers RHQG and G-UA9M.

Acknowledgments: The authors gratefully appreciate and acknowledge the financial support from the Hong Kong Polytechnic University.

Conflicts of Interest: The authors declare no conflict of interest.

References

1. Yablonovitch, E. Inhibited spontaneous emission in solid-state physics and electronics. *Phys. Rev. Lett.* **1987**, *58*, 2059–2062. [CrossRef] [PubMed]
2. John, S. Strong localization of photons in certain disordered dielectric superlattices. *Phys. Rev. Lett.* **1987**, *58*, 2486–2489. [CrossRef] [PubMed]
3. Yablonovitch, E.; Gmitter, T.J. Photonic band structure: The face-centered-cubic case. *Phys. Rev. Lett.* **1989**, *63*, 1950. [CrossRef] [PubMed]
4. Joannopoulos, J.D.; Villeneuve, P.R.; Fan, S. Photonic crystals: Putting a new twist on light. *Nature* **1997**, *386*, 143. [CrossRef]
5. Zhang, J.; Sun, Z.; Yang, B. Self-assembly of photonic crystals from polymer colloids. *Curr. Opin. Colloid Interface Sci.* **2009**, *14*, 103–114. [CrossRef]
6. Paquet, C.; Kumacheva, E. Nanostructured polymers for photonics. *Mater. Today* **2008**, *11*, 48–56. [CrossRef]
7. Wang, W.; Tang, B.; Ma, W.; Zhang, J.; Ju, B.; Zhang, S. Easy approach to assembling a biomimetic color film with tunable structural colors. *J. Opt. Soc. Am. A Opt. Image Sci. Vis.* **2015**, *32*, 1109–1117. [CrossRef] [PubMed]
8. Gu, Z.; Uetsuka, H.; Takahashi, K.; Nakajima, R.; Onishi, H.; Fujishima, A.; Stao, O. Structural color and the lotus effect. *Angew. Chem. Int. Ed.* **2003**, *42*, 894–897. [CrossRef] [PubMed]
9. Lu, Y.; Yin, Y.; Li, Z.; Xia, Y. Synthesis and self-assembly of au@ sio2 core-shell colloids. *Nano Lett.* **2002**, *2*, 785–788. [CrossRef]
10. Gu, Z.Z.; Chen, H.; Zhang, S.; Sun, L.; Xie, Z.; Ge, Y. Rapid synthesis of monodisperse polymer spheres for self-assembled photonic crystals. *Colloids Surfaces A Physicochem. Eng. Asp.* **2007**, *302*, 312–319. [CrossRef]
11. Egen, M.; Zentel, R. Surfactant-free emulsion polymerization of various methacrylates: Towards monodisperse colloids for polymer opals. *Macromol. Chem. Phys.* **2004**, *205*, 1479–1488. [CrossRef]
12. Yuan, W.; Zhang, K.Q. Structural evolution of electrospun composite fibers from the blend of polyvinyl alcohol and polymer nanoparticles. *Langmuir* **2012**, *28*, 15418–15424. [CrossRef] [PubMed]
13. Ha, S.T.; Park, O.O.; Im, S.H. Size control of highly monodisperse polystyrene particles by modified dispersion polymerization. *Macromol. Res.* **2010**, *18*, 935–943. [CrossRef]
14. Wang, J.; Wen, Y.; Ge, H.; Sun, Z.; Zheng, Y.; Song, Y.; Jiang, L. Simple fabrication of full color colloidal crystal films with tough mechanical strength. *Macromol. Chem. Phys.* **2006**, *207*, 596–604. [CrossRef]
15. Ramakrishna, S. *An Introduction to Electrospinning and Nanofibers*; World Scientific: Singapore, 2005; p. 15, ISBN 981-256-415-2.
16. Mu, Q.; Zhang, Q.; Gao, L.; Chu, Z.; Cai, Z.; Zhang, X.; Wang, K.; Wei, Y. Structural evolution and formation mechanism of the soft colloidal arrays in the core of paam nanofibers by electrospun packing. *Langmuir* **2017**, *33*, 10291–10301. [CrossRef] [PubMed]
17. Bhardwaj, N.; Kundu, S.C. Electrospinning: A fascinating fiber fabrication technique. *Biotechnol. Adv.* **2010**, *28*, 325–347. [CrossRef] [PubMed]
18. Zhang, J.; He, S.; Liu, L.; Guan, G.; Lu, X.; Sun, X.; Peng, H. The continuous fabrication of mechanochromic fibers. *J. Mater. Chem. C* **2016**, *4*, 2127–2133. [CrossRef]
19. Ruhl, T.; Spahn, P.; Hellmann, G.P. Artificial opals prepared by melt compression. *Polymer* **2003**, *44*, 7625–7634. [CrossRef]
20. Finlayson, C.E.; Spahn, P.; Snoswell, D.R.; Yates, G.; Kontogeorgos, A.; Haines, A.I.; Hellmann, G.P.; Baumberg, J.J. 3d bulk ordering in macroscopic solid opaline films by edge-induced rotational shearing. *Adv. Mater.* **2011**, *23*, 1540–1544. [CrossRef] [PubMed]
21. Pursiainen, O.L.J.; Baumberg, J.J.; Winkler, H.; Viel, B.; Spahn, P.; Ruhl, T. Shear-induced organization in flexible polymer opals. *Adv. Mater.* **2008**, *20*, 1484–1487. [CrossRef]

22. Kohri, M.; Yanagimoto, K.; Kawamura, A.; Hamada, K.; Imai, Y.; Watanabe, T.; Ono, T.; Taniguchi, T.; Kishikawa, K. Polydopamine-based 3d colloidal photonic materials: Structural color balls and fibers from melanin-like particles with polydopamine shell layers. *ACS Appl Mater. Interfaces* **2018**, *10*, 7640–7648. [CrossRef] [PubMed]
23. Giersig, M.; Mulvaney, P. Preparation of ordered colloid monolayers by electrophoretic deposition. *Langmuir* **1993**, *9*, 3408–3413. [CrossRef]
24. Yu, H.; Liao, D.; Johnston, M.B.; Li, B. All-optical full-color displays using polymer nanofibers. *ACS Nano* **2011**, *5*, 2020–2025. [CrossRef] [PubMed]
25. Liu, Z.; Zhang, Q.; Wang, H.; Li, Y. Structurally colored carbon fibers with controlled optical properties prepared by a fast and continuous electrophoretic deposition method. *Nanoscale* **2013**, *5*, 6917–6922. [CrossRef] [PubMed]
26. Lim, J.M.; Moon, J.H.; Yi, G.R.; Heo, C.J.; Yang, S.M. Fabrication of one-dimensional colloidal assemblies from electrospun nanofibers. *Langmuir* **2006**, *22*, 3445–3449. [CrossRef] [PubMed]
27. Liu, Z.; Zhang, Q.; Wang, H.; Li, Y. Structural colored fiber fabricated by a facile colloid self-assembly method in micro-space. *Chem. Commun. (Camb.)* **2011**, *47*, 12801–12803. [CrossRef] [PubMed]
28. Josephson, D.P.; Miller, M.; Stein, A. Inverse opal SiO_2 photonic crystals as structurally-colored pigments with additive primary colors. *Z. Anorg. Allg. Chem.* **2014**, *640*, 655–662. [CrossRef]
29. Katritzky, A.R.; Sild, S.; Karelson, M. Correlation and prediction of the refractive indices of polymers by qspr. *J. Chem. Inf. Comput. Sci.* **1998**, *38*, 1171–1176. [CrossRef]
30. Jicerano, J. *Prediction of Polymer Properties*; CRC Press: New York, NY, USA, 2002; p. 272, ISBN 0-8247-0821-0.
31. Sato, O.; Kubo, S.; Gu, Z.Z. Structural color films with lotus effects, superhydrophilicity, and tunable stop-bands. *Acc. Chem. Res.* **2008**, *42*, 1–10. [CrossRef] [PubMed]
32. Jin, Y.; Yang, D.; Kang, D.; Jiang, X. Fabrication of necklace-like structures via electrospinning. *Langmuir* **2010**, *26*, 1186–1190. [CrossRef] [PubMed]
33. Crespy, D.; Friedemann, K.; Popa, A.M. Colloid-electrospinning: Fabrication of multicompartment nanofibers by the electrospinning of organic or/and inorganic dispersions and emulsions. *Macromol. Rapid Commun.* **2012**, *33*, 1978–1995. [CrossRef] [PubMed]
34. Zhang, C.L.; Yu, S.H. Nanoparticles meet electrospinning: Recent advances and future prospects. *Chem. Soc. Rev.* **2014**, *43*, 4423–4448. [CrossRef] [PubMed]
35. Dzenis, Y. Spinning continuous fibers for nanotechnology. *Science* **2004**, *304*, 1917–1919. [CrossRef] [PubMed]
36. Zhao, J.; Liu, H.; Xu, L. Preparation and formation mechanism of highly aligned electrospun nanofibers using a modified parallel electrode method. *Mater. Des.* **2016**, *90*, 1–6. [CrossRef]
37. Li, D.; Wang, Y.L.; Xia, Y.N. Electrospinning of polymeric and ceramic nanofibers as uniaxially aligned arrays. *Nano Lett.* **2003**, *3*, 1167–1171. [CrossRef]
38. Dersch, R.; Liu, T.; Schaper, A.K.; Greiner, A.; Wendorff, J.H. Electrospun nanofibers: Internal structure and intrinsic orientation. *Polym. Chem.* **2003**, *41*, 545–553. [CrossRef]
39. Teo, W.E.; Ramakrishna, S. A review on electrospinning design and nanofibre assemblies. *Nanotechnology* **2006**, *17*, R89–R106. [CrossRef] [PubMed]
40. Li, D.; Wang, Y.; Xia, Y. Electrospinning nanofibers as uniaxially aligned arrays and layer-by-layer stacked films. *Adv. Mater.* **2004**, *16*, 361–366. [CrossRef]
41. Eablonovitch, E. Photonic band-gap structures. *J. Opt. Soc. Am. B* **1993**, *10*, 283–295. [CrossRef]
42. Lee, C.H.; Yu, J.; Wang, Y.; Tang, A.Y.L.; Kan, C.W.; Xin, J.H. Effect of graphene oxide inclusion on the optical reflection of a silica photonic crystal film. *RSC Adv.* **2018**, *8*, 16593–16602. [CrossRef]
43. Gao, W.; Rigout, M.; Owens, H. Self-assembly of silica colloidal crystal thin films with tuneable structural colours over a wide visible spectrum. *Appl. Surf. Sci.* **2016**, *380*, 12–15. [CrossRef]
44. Mallakpour, S.; Behranvand, V. Polymeric nanoparticles: Recent development in synthesis and application. *Express Polym. Lett.* **2016**, *10*, 895–913. [CrossRef]
45. Iler, R.K. Multilayers of colloidal particles. *J. Colloid Interface Sci.* **1996**, *21*, 569–594. [CrossRef]
46. Stöber, W.; Fink, A.; Bohn, E. Controlled growth of monodisperse silica spheres in the micron size range. *J. Colloid Interface Sci.* **1968**, *26*, 62–69. [CrossRef]
47. Li, Q.; Zhang, Y.; Shi, L.; Qiu, H.; Zhang, S.; Qi, N.; Hu, J.; Yuan, W.; Zhang, X.; Zhang, K.Q. Additive mixing and conformal coating of noniridescent structural colors with robust mechanical properties fabricated by atomization deposition. *ACS Nano* **2018**, *12*, 3095–3102. [CrossRef] [PubMed]

48. Sopyan, I.; Watanabe, M.; Murasawa, S.; Hashimoto, K.; Fujishima, A. Efficient TiO$_2$ powder and film photocatalysts with rutile crystal structure. *Chem. Lett.* **1996**, *25*, 69–70. [CrossRef]
49. Chrysicopoulou, P.; Davazogloub, D.; Trapalis, C.; Kordasa, G. Optical properties of very thin (100 nm) sol–gel TiO$_2$ films. *Thin Solid Films* **1998**, *323*, 188–193. [CrossRef]
50. Yuan, X.; Xu, W.; Huang, F.; Chen, D.; Wei, Q. Structural colour of polyester fabric coated with ag/tio2 multilayer films. *Surf. Eng.* **2016**, *33*, 231–236. [CrossRef]
51. Guo, D.; Ito, A.; Goto, T.; Tu, R.; Wang, C.; Shen, Q.; Zhang, L. Effect of laser power on orientation and microstructure of TiO$_2$ films prepared by laser chemical vapor deposition method. *Mater. Lett.* **2013**, *93*, 179–182. [CrossRef]
52. Sun, S.Q.S.B.; Zhang, W.Q.; Wang, D. Preparation and antibacterial activity of ag-tio 2 composite film by liquid phase deposition (lpd) method. *Bull. Mater. Sci.* **2008**, *31*, 61–66. [CrossRef]
53. Herbig, B.; Löbmann, P. TiO$_2$ photocatalysts deposited on fiber substrates by liquid phase deposition. *J. Photochem. Photobiol. A Chem.* **2004**, *163*, 359–365. [CrossRef]
54. Chen, F.; Yang, H.; Li, K.; Deng, B.; Li, Q.; Liu, X.; Dong, B.; Xiao, X.; Wang, D.; Qin, Y.; et al. Facile and effective coloration of dye-inert carbon fiber fabrics with tunable colors and excellent laundering durability. *ACS Nano* **2017**, *11*, 10330–10336. [CrossRef] [PubMed]
55. Zeng, J.; Huang, J.; Lu, W.; Wang, X.; Wang, B.; Zhang, S.; Hou, J. Necklace-like noble-metal hollow nanoparticle chains: Synthesis and tunable optical properties. *Adv. Mater.* **2007**, *19*, 2172–2176. [CrossRef]
56. Haynes, C.L.; Van Duyne, R. Nanosphere lithography: A versatile nanofabrication tool for studies of size-dependent nanoparticle optics. *J. Phys. Chem. B* **2001**, *105*, 5599–5611. [CrossRef]
57. Luo, Y.; Zhang, J.; Sun, A.; Chu, C.; Zhou, S.; Guo, J.; Chen, T.; Xu, G. Electric field induced structural color changes of sio2@tio2 core–shell colloidal suspensions. *J. Mater. Chem. C* **2014**, *2*, 1990–1994. [CrossRef]
58. Vanderhoff, J.W.; Vitkuske, J.F.; Bradford, E.B.; Alfrey, T., Jr. Some factors involved in the preparation of uniform particle size latexes. *J. Polym. Sci. Part A Polym. Chem.* **1956**, *20*, 225–234. [CrossRef]
59. Kim, S.H.; Lee, S.Y.; Yang, S.M.; Yi, G.R. Self-assembled colloidal structures for photonics. *NPG Asia Mater.* **2011**, *3*, 25–33. [CrossRef]
60. Meng, Y.; Tang, B.; Xiu, J.; Zheng, X.; Ma, W.; Ju, B.; Zhang, S. Simple fabrication of colloidal crystal structural color films with good mechanical stability and high hydrophobicity. *Dyes Pigments* **2015**, *123*, 420–426. [CrossRef]
61. Han, M.G.; Heo, C.-J.; Shim, H.; Shin, C.G.; Lim, S.-J.; Kim, J.W.; Jin, Y.W.; Lee, S. Structural color manipulation using tunable photonic crystals with enhanced switching reliability. *Adv. Opt. Mater.* **2014**, *2*, 535–541. [CrossRef]
62. Tang, B.; Xu, Y.; Lin, T.; Zhang, S. Polymer opal with brilliant structural color under natural light and white environment. *J. Mater. Res.* **2015**, *30*, 3134–3141. [CrossRef]
63. Tang, B.; Zheng, X.; Lin, T.; Zhang, S. Hydrophobic structural color films with bright color and tunable stop-bands. *Dyes Pigments* **2014**, *104*, 146–150. [CrossRef]
64. Tanrisever, T.; Okay, O.; Soenmezoğlu, I.C. Kinetics of emulsifier-free emulsion polymerization of methyl methacrylate. *J. Appl. Polym. Sci.* **1996**, *61*, 485–493. [CrossRef]
65. Zou, D.; Ma, S.; Guan, R.; Park, M.; Sun, L.; Aklonis, J.J.; Salovey, R. Model filled polymers. V. Synthesis of crosslinked monodisperse polymethacrylate beads. *J. Polym. Sci. Part A Polym. Chem.* **1992**, *30*, 137–144. [CrossRef]
66. Tang, B.; Wu, C.; Lin, T.; Zhang, S. Heat-resistant pmma photonic crystal films with bright structural color. *Dyes Pigments* **2013**, *99*, 1022–1028. [CrossRef]
67. Park, J.G.; Kim, S.H.; Magkiriadou, S.; Choi, T.M.; Kim, Y.S.; Manoharan, V.N. Full-spectrum photonic pigments with non-iridescent structural colors through colloidal assembly. *Angew. Chem. Int. Ed. Engl.* **2014**, *53*, 2899–2903. [CrossRef] [PubMed]
68. Liu, G.; Zhou, L.; Wang, C.; Wu, Y.; Li, Y.; Fan, Q.; Shao, J. Study on the high hydrophobicity and its possible mechanism of textile fabric with structural colors of three-dimensional poly(styrene-methacrylic acid) photonic crystals. *RSC Adv.* **2015**, *5*, 62855–62863. [CrossRef]
69. Yuan, W.; Zhou, N.; Shi, L.; Zhang, K.Q. Structural coloration of colloidal fiber by photonic band gap and resonant mie scattering. *ACS Appl. Mater. Interfaces* **2015**, *7*, 14064–14071. [CrossRef] [PubMed]
70. Jia, Y.; Zhang, Y.; Zhou, Q.; Fan, Q.; Shao, J. Structural colors of the SiO$_2$/polyethyleneimine thin films on poly(ethylene terephthalate) substrates. *Thin Solid Films* **2014**, *569*, 10–16. [CrossRef]

71. Liu, G.; Zhou, L.; Zhang, G.; Li, Y.; Chai, L.; Fan, Q.; Shao, J. Fabrication of patterned photonic crystals with brilliant structural colors on fabric substrates using ink-jet printing technology. *Mater. Des.* **2017**, *114*, 10–17. [CrossRef]
72. Li, Y.; Zhou, L.; Liu, G.; Chai, L.; Fan, Q.; Shao, J. Study on the fabrication of composite photonic crystals with high structural stability by co-sedimentation self-assembly on fabric substrates. *Appl. Surf. Sci.* **2018**, *444*, 145–153. [CrossRef]
73. Huang, Z.-M.; Zhang, Y.Z.; Kotaki, M.; Ramakrishna, S. A review on polymer nanofibers by electrospinning and their applications in nanocomposites. *Compos. Sci. Technol.* **2003**, *63*, 2223–2253. [CrossRef]
74. Zhang, F.; Ma, X.; Cao, C.; Li, J.; Zhu, Y. Poly(vinylidene fluoride)/SiO$_2$ composite membranes prepared by electrospinning and their excellent properties for nonwoven separators for lithium-ion batteries. *J. Power Sources* **2014**, *251*, 423–431. [CrossRef]
75. Im, J.S.; Kim, M.I.; Lee, Y.S. Preparation of pan-based electrospun nanofiber webs containing TiO$_2$ for photocatalytic degradation. *Mater. Lett.* **2008**, *62*, 3652–3655. [CrossRef]
76. Pant, H.R.; Pandeya, D.R.; Nam, K.T.; Baek, W.I.; Hong, S.T.; Kim, H.Y. Photocatalytic and antibacterial properties of a TiO$_2$/nylon-6 electrospun nanocomposite mat containing silver nanoparticles. *J. Hazard. Mater.* **2011**, *189*, 465–471. [CrossRef] [PubMed]
77. Kanehata, M.; Ding, B.; Shiratori, S. Nanoporous ultra-high specific surface inorganic fibres. *Nanotechnology* **2007**, *18*, 315602. [CrossRef]
78. Lim, J.M.; Yi, G.R.; Moon, J.H.; Heo, C.J.; Yang, S.M. Superhydrophobic films of electrospun fibers with multiple-scale surface morphology. *Langmuir* **2007**, *23*, 7981–7989. [CrossRef] [PubMed]
79. Stoiljkovic, A.; Ishaque, M.; Justus, U.; Hamel, L.; Klimov, E.; Heckmann, W.; Eckhardt, B.; Wendorff, J.H.; Greiner, A. Preparation of water-stable submicron fibers from aqueous latex dispersion of water-insoluble polymers by electrospinning. *Polymer* **2007**, *48*, 3974–3981. [CrossRef]

© 2018 by the authors. Licensee MDPI, Basel, Switzerland. This article is an open access article distributed under the terms and conditions of the Creative Commons Attribution (CC BY) license (http://creativecommons.org/licenses/by/4.0/).

MDPI
St. Alban-Anlage 66
4052 Basel
Switzerland
Tel. +41 61 683 77 34
Fax +41 61 302 89 18
www.mdpi.com

Fibers Editorial Office
E-mail: fibers@mdpi.com
www.mdpi.com/journal/fibers

www.ingramcontent.com/pod-product-compliance
Lightning Source LLC
LaVergne TN
LVHW071948080526
838202LV00064B/6708